Experimental Oceanography

by

Sam Kelly

HOWARD W. SAMS & CO., INC.
THE BOBBS-MERRILL CO., INC.
INDIANAPOLIS · KANSAS CITY · NEW YORK

FIRST EDITION

FIRST PRINTING—1975

International Standard Book Number: 0-672-21307-9
Library of Congress Catalog Card Number: 75-28964

Preface

No natural wonder holds as much fascination for man as the oceans. Few can watch the sea and not be challenged to investigate its secrets. The oceans are justly called man's last frontier on earth. Even with our sophisticated space age technology, our explorations have been limited to only a tiny portion of the hydrospace.

Since the sea resists direct observation, man must depend upon instruments for his studies. This book is intended to be a "how to" book. In it you will learn how to build some of the basic instruments and tools of the oceanographer. With these, you will be able to study some of the secrets of the seas.

As a result of your research with these instruments, you may become so fascinated with the study of the oceans that you will want to learn more about them. Possibly you will make their study your career. This could be a wise choice since few vocations offer the challenge, excitement, and potential for discovery that oceanography does.

SAM KELLY

Contents

CHAPTER 1

INTRODUCTION 7
The Nansen Bottle — The Reversing Thermometer — SONAR — Echo Sounding — The Bathythermograph — Diving — Unmanned Remote-Controlled Underwater Vehicles — Manned Submersibles

CHAPTER 2

BASIC CONCEPTS 37
Topography — Waves, Tides, and Currents

CHAPTER 3

UNITS OF MEASUREMENT 47

CHAPTER 4

CONSTRUCTION PRACTICES 51

CHAPTER 5

INSTRUMENT AND APPARATUS PROJECTS 57
Electronic Thermometer — Current Speed and Direction Measuring System — Wind Speed and Direction Indicators — Transparency and the Secchi Disc — Underwater Sound System — An Aquatic Biotelemetry System — Plankton Sampling Net

BIBLIOGRAPHY 91

INDEX 93

1

Introduction

What is Oceanography? The word means many things to different people. We will define oceanography as a complex science dedicated to the studies of the oceans. It requires the services of many fields of scientific endeavor. In the past, it has been classically divided into three disciplines: physical, chemical, and biological. These divisions have become too simplified for the twentieth century. Modern oceanography requires a more diversified team and equipment (Figs. 1-1 and 1-2).

Most of our knowledge of the oceans depends on instrumentation rather than on direct observation. Therefore, the services of engineers are vital for designing instruments and monitoring systems. Even the most modest oceanographic survey results in a tremendous volume of data. This mountain of raw numbers can be efficiently transformed into meaningful information only by modern digital computers. A data processing specialist, usually an electronics engineer skilled in computer programming, is indispensable. In addition, chemists, biologists, meteorologists, physical oceanographers, geologists, mathematicians, and physicists are necessary to complete the scientific team.

Behind the engineers and scientists are a wide range of support workers ranging from skilled carpenters to artists. Chart 1-1 lists the variety of skills and crafts that are required to make up a successful oceanographic expedition. Admittedly, it is not complete.

In addition to the scientific team, there is another group of workers equally vital to the success of the expedition. This group includes the vessel's crew, navigator, captain, ship's chandler, and longshoremen. Oceanography is truly a team science. It would be

Courtesy National Oceanic and Atmospheric Administration

Fig. 1-1. An oceangoing research ship.

impossible to name a single specialty as being the most important for success of an expedition. However, the ship's cook would probably rate high if the subject was brought to a vote!

Fig. 1-2. A typical inshore research vessel.

Chart 1-1. Oceanographic Team Members

SCIENTISTS	
Physical Oceanographers	Geologists
Biologists	Meteorologists
Chemists	Bacteriologists
Zoologists	Physicists
ENGINEERS	
Electronics—	Mechanical—
Instrumentation	Instrumentation
Data processing (computers)	Structural (ocean engineering)
Communications	Ship control systems
Video (underwater television)	Buoy technology
Space technology (remote sensing)	Civil—
Marine Architecture	Wave effects
Photogrammetry—	Mass transport of pollutants
Still and motion pictures	
Multispectral sensors	
SKILLED CRAFTSMEN AND TECHNICIANS	
Carpenters	Water-quality technicians
Marine plumbers	Mechanics
Artists	Photographers
Electronic technicians	Divers

Beyond the altruistic observation that our inner space is the last unexplored frontier on our planet lies the problem of diminishing natural resources. In the decade to come, man may find that his knowledge of the oceans spells the difference between survival and extinction of one more species of animal—man!

Man's earliest studies of the oceans were concerned mainly with the furtherance of trade. Soundings and observations concerning currents, waves, and tides were of greatest interest. Among the earliest reported measurements was the sounding of a depth of over 1800 meters in the sea of Sardinia two centuries before Christ!

It wasn't until the nineteenth century that interest in oceanic studies became more scientifically oriented. One of the men responsible for great scientific advancement during this period was Lieutenant Matthew Fontaine Maury of the United States Navy.

Lieutenant Maury served as officer in charge of the U.S. Navy's depot of charts and instruments in Washington, D.C. from 1842 to 1861. The main purpose of the depot was to serve as a storehouse for maps and charts. Maury was not content with what could have been a comfortable job as a storekeeper. He originated a system in which ship captains would mark their observations on nautical charts and send these annotated charts to the depot. Maury then incorporated their comments on new charts and returned a copy to the ship's captain without charge.

He went even further and organized ocean surveys and printed important navigation texts such as Nathaniel Bowditch's *The New American Practical Navigator* and special sailing instructions. He continued Benjamin Franklin's studies of the Gulf Stream, coining the name "the river in the ocean" for it. He did extensive charting of the Gulf Stream and provided information showing how its course varied with the seasons. As a result of his hydrographic and cartographic work, the passage from England to California was shortened by thirty days!

It was through Maury's efforts that the Brussels convention of 1853 was convened. At this conference, representatives from the leading maritime nations planned a uniform program for reporting oceanographic information.

He was responsible for the preparation of the first bathymetric charts. These were charts that showed the bottom features with contour lines as deep as 4000 fathoms (a fathom is equal to six feet). In conjunction with this work, he had a large number of bottom sediment samples analyzed.

With the outbreak of civil war in the United States, Maury resigned his commission in the United States Navy and returned to his native Virginia where he served with distinction in the Confederate States Navy.

The next major advance in oceanography was brought about by the British Navy. Instrumentation had advanced considerably in the period from 1850 to 1870. The British empire was the unquestioned military and naval power in the world at that time. Therefore, it was logical that they should undertake an oceanographic expedition with a well-equipped vessel. In 1872, the first major oceanographic expedition got under way. This was the famous Challenger expedition which lasted four years. The cataloging of specimens and reduction of oceanographic data from this expedition required an additional thirty years! One of the most significant findings was that the proportion of salts in seawater remains constant, regardless of what part of the ocean the sample is taken in or of the overall salinity.

Toward the end of the nineteenth century, other governments and private individuals sponsored expeditions. The prominence of wealthy sponsors and participants gave the name "the gentleman's science" to oceanography.

After a period of prominence, oceanography took a backseat to other, more glamorous, fields of science. During World War I, oceanographers were called upon to help with antisubmarine warfare. In World War II, they were called upon in planning the amphibious campaigns. With the public's realization of the environmental and natural resource problems, the science has emerged to new prominence.

Instrumentation plays an overwhelmingly important part in the study of our oceans.

THE NANSEN BOTTLE

Getting samples of the oceanic waters at varying depths is no easy task. The most significant advance along these lines was the

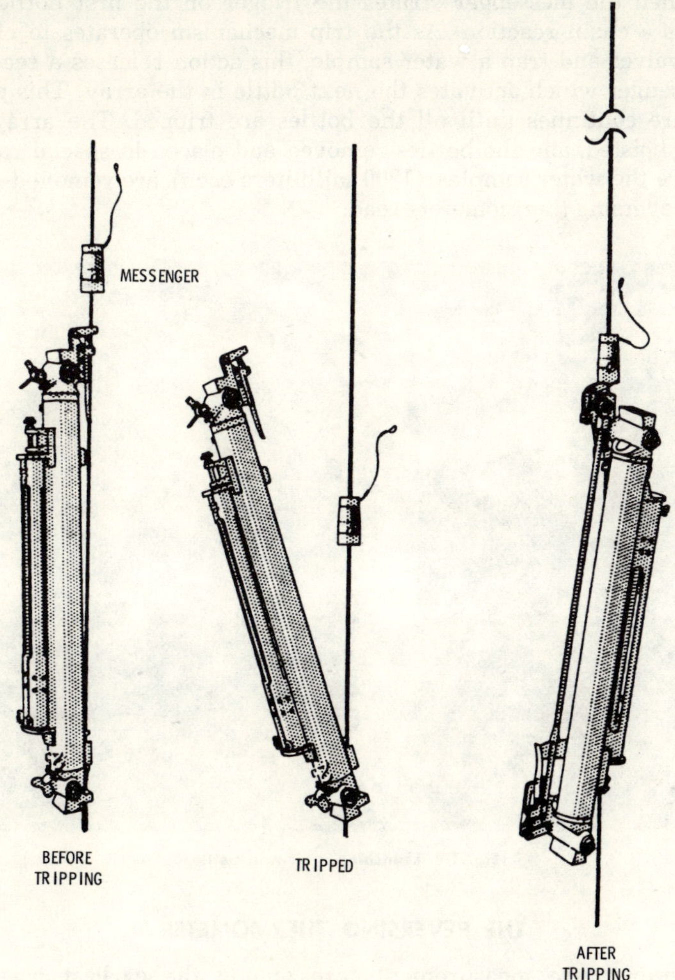

Fig. 1-3. Nansen bottle operation.

invention of a sampling bottle by Fridtjof Nansen, a Norwegian scientist and polar explorer. The "bottle" consists of a brass tube with valves at each end, with a mechanism for reversing the bottle and tripping both valves when a metal weight called a messenger strikes a trigger. Protective cages for reversing thermometers are attached to the side (Fig. 1-3). In practice, these bottles (now called Nansen bottles) are clamped at intervals on a wire, then lowered into the sea (Fig. 1-4). When the array reaches the desired depth, the messenger, a circular weight, is slipped over the wire and dropped.

When the messenger strikes the trigger on the first bottle, it starts a chain reaction. As the trip mechanism operates to close the valves and trap a water sample, this action releases a second messenger which activates the next bottle in the array. This procedure continues until all the bottles are tripped. The array is then hoisted, and the bottles removed and placed in special racks where the water samples (1200 milliliters each) are removed and the reversing thermometers read.

Fig. 1-4. Launching a Nansen bottle.

THE REVERSING THERMOMETER

Temperature measurements are among the earliest oceanographic measurements taken and constitute the greatest volume of data compiled to date. The two basic temperature-measuring devices are the bucket thermometer and the reversing thermometer. The bucket thermometer, as the name implies, is simply a bucket of surface water that is hauled aboard, and then its temperature is measured with a mercury thermometer. The reversing thermometer registers the temperature at the moment it is inverted.

Subsurface thermometers have typically been made with reversing thermometers. Even today, the reversing thermometer (Fig. 1-5), though delicate, remains the most reliable and ac-

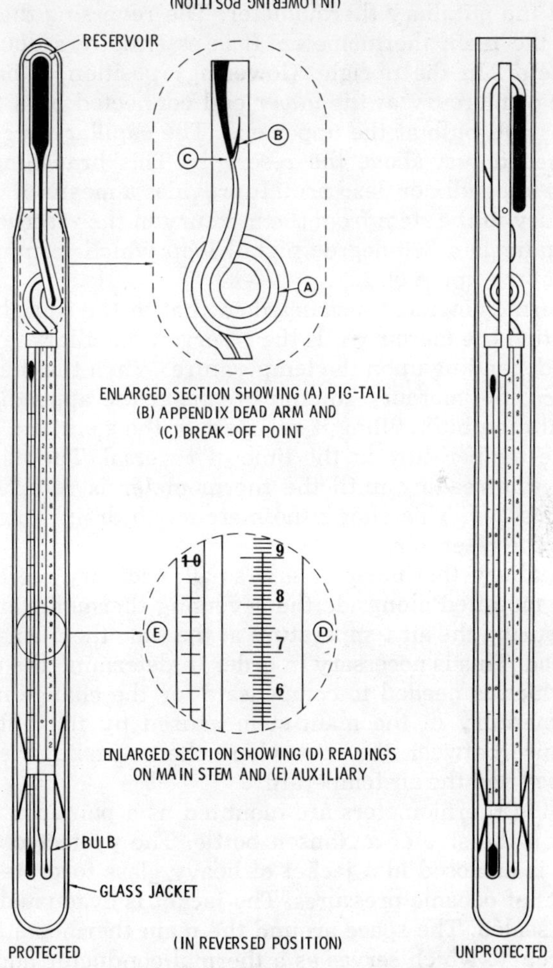

Fig. 1-5. Deep-sea reversing thermometers.

curate means of measuring sea temperatures. Good-quality reversing thermometers measure temperatures to ±0.02°C, and depths to ±0.5% of the actual value. Special reversing thermometers are accurate to ±0.005°C. Their limitations are that they provide temperature only for discrete points and must be brought on board to be read.

There are two types of reversing thermometers, protected and unprotected. The Celsius scale is universally used. The thermometers are hand calibrated by the manufacturers and are periodically recalibrated during their life.

Each instrument actually consists of two thermometers, the main and the auxiliary thermometer. The reversing thermometer is called the main thermometer. It is essentially a double-ended thermometer. In the upright (lowering) position, it has a large reservoir of mercury at the *lower* end connected by a fine capillary to a small bulb at the upper end. The capillary is constricted and branched just above the reservoir. This branching point is called the appendix or dead arm. It provides a means of separating the mercury in the stem from the mercury in the reservoir. Above the appendix is a 360-degree pigtail loop which terminates with the bulb at the upper end.

The thermometer is constructed so that in the upright (lowering) position the mercury fills the reservoir, capillary, and part of the bulb depending upon the temperature. When the thermometer is reversed, the mercury column breaks at the appendix and descends into the bulb, filling it and part of the stem and thus indicating the temperature at the time of reversal. The mercury remains at this reading until the thermometer is returned to the upright position. This allows the mercury to drain from the bulb back into the reservoir.

The auxiliary thermometer is a small, ordinary mercury thermometer mounted alongside the reversing thermometer to facilitate measuring the air temperature at the time the main thermometer is read. This is necessary in order to determine the correction factor, which is needed to compensate for the change in the volume of mercury in the main stem caused by the difference in temperature between the spot where the reversing thermometer was tripped and the air temperature.

Reversing thermometers are mounted as a pair in a protective housing on the side of a Nansen bottle. The protected thermometer pair is enclosed in a jacket of heavy glass to protect it from the effects of oceanic pressures. The jacket is evacuated and both ends are sealed. The space around the main thermometer is filled with mercury, which serves as a thermal conductor and provides greater sensitivity to the temperature changes. Protection from the hydrostatic pressure provides a true reading of water temperature *in situ*.

Mounted alongside the protected thermometer is one open to the hydrostatic pressure, and aptly named the unprotected thermometer. It is in a thick glass jacket that is open at the top. Therefore, it does not give a true temperature reading but one that is increased approximately 0.1°C for each 100 meters of depth. The unprotected thermometer can be used as a pressure gauge for determining the exact depth at which the thermometers were reversed.

SONAR

The term *sonar* is most frequently associated with antisubmarine warfare. To be sure, the concept originated with warfare. During the closing months of World War I, the Allied Submarine Devices Investigating Committee started work on an advanced hydrophone, or underwater microphone, for locating submarines. The committee disbanded after the war, but the British pursued the science and developed an improved locating device. They named this invention ASDIC after the committee. United States developments resulted in an improved system that was named sonar (*SO*und *NA*vigation and *R*anging).

Sonar is an electronic device that uses sound energy as a means of locating submerged objects such as wrecks, submarines, etc. Most sonar systems are designed so that they can be used in both an active and passive mode. In the active mode, the system generates a powerful pulse of sound energy. This energy is transformed into a pressure wave by the transducer. This pulse strikes the target and is reflected back in the form of an echo. The transducer acts to receive the reflected echo and convert the pressure back into electrical energy. This weak electrical signal is amplified by the receiver and displayed on an indicator. The most commonly encountered indicator is the cathode-ray oscilloscope. The distance to the target is determined by the time required for the pulse to strike the target and return, and the knowledge of the speed of sound through seawater.

The early sonar sets utilized a transducer that had to be rotated or trained to a particular bearing in order to transmit and receive on that bearing. This was called the searchlight sonar. More modern sonars transmit in a 360° beam, and the azimuth to the target is determined by an electronic beam scanning process rather than by a mechanically rotated transducer.

In the passive, or hydrophone, mode, the transmitter is turned off and the sensitive receiver is tuned to receive sounds from targets that generate sound waves. These include marine animals such as whales, porpoises, snapping shrimp, croakers, and others. Other sources of sound in the ocean are ship's machinery and rain. The small hydrophone described in Chapter 5 will allow you to explore this secret world of underwater sound.

The sonar principle is used in the echo sounder. The primary difference is that the echo sounder's transducer is fixed in the hull of the ship looking downward and that the data display presents the range to the bottom only.

ECHO SOUNDING

The single instrument that has resulted in the greatest saving in measurement time at sea is the echo sounder. Soundings are the most ancient of oceanographic observations. Since man began using the oceans for commerce, a major cause of marine disaster has been running aground or striking submerged rocks and pinnacles. The lead line, or hand line, has historically been used for soundings. This device is simply a measured and marked line with a lead weight attached to it. For deep ocean measurements, cannonballs were often used. The ship had to be stopped to make accurate soundings. In the late nineteenth century, Lord Kelvin of England perfected a mechanical depth-sounder.

However, sounding was still a tedious time-consuming process. In the early twentieth century Alexander Behm, a German, developed the theory of acoustic sounding using underwater explosives. In 1919 the United States Navy's research laboratory, the forerunner of the current Naval Electronics Laboratory, produced the first practical electronic echo sounder. A pioneer company in producing commercial echo sounders was the Submarine Signal Company, now part of the Raytheon Corporation.

The echo sounder, Fig. 1-6, is an electronic instrument that determines water depth by measuring the time required for a pulse of acoustic energy to travel from the surface to the bottom and back.

The echo sounder system consists of a transducer, to transmit and receive the acoustic pulses; a transmitter, to generate and time the pulses; a receiver, to detect and amplify the weak echo; and an indicator, to display the information. In operation, the transmitter generates a pulse or short burst of energy. The transducer converts this electrical energy into an acoustic pressure wave directed downward. This wave strikes the ocean floor. Part of the energy is reflected back. The transducer receives this echo. It is then detected and amplified by the receiver. The output of the receiver is displayed in a variety of ways. In low-cost echo sounders a flashing, rotating neon indicator is most commonly encountered. More elaborate instruments, such as those found on fishing vessels and commercial ships, employ cathode-ray-tube displays and strip-chart recorders. Precision depth recorders, such as those used on oceanographic survey ships, use high-resolution electrographic recorders. An electrically sensitive paper is the recording medium. Frequently these precision echo sounders employ elaborate filters and special coding circuitry to enable the operator to eliminate unwanted echoes and to allow precise scaling of the charts.

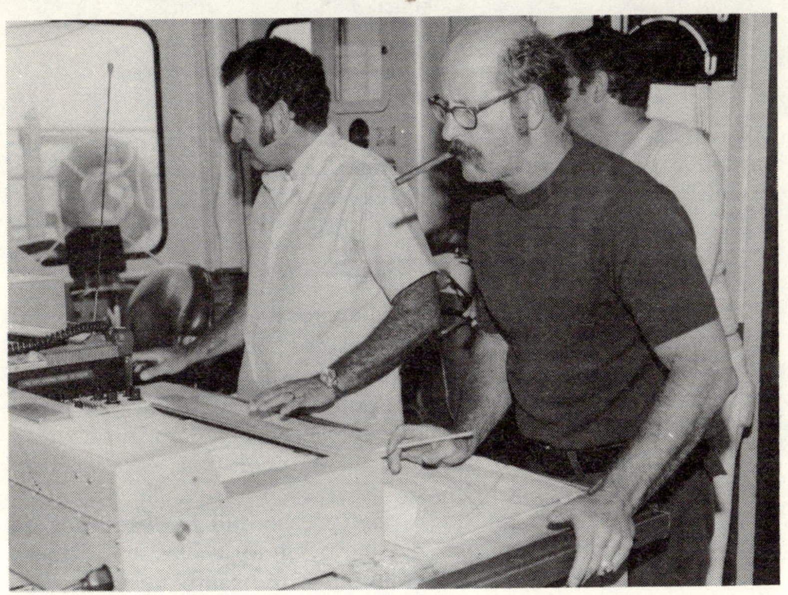

Fig. 1-6. Operating a highly precise echo sounder.

For accurate depth measurement with an echo sounder, the speed of sound through the water must be known precisely. Sound travels in air at a speed of approximately 1100 feet per second. In water it travels at approximately 4800 feet per second. Fig. 1-7 illustrates how the speed of sound varies with depth as a function of pressure, salinity, and temperature.

The echo sounder is fully automatic and can be operated while the ship is underway. This permits it to be used as a navigation aid by comparing the echo sounder record with soundings on navigation charts. It will also reveal uncharted shoals and pinnacles, thus greatly increasing the safety of ship operations.

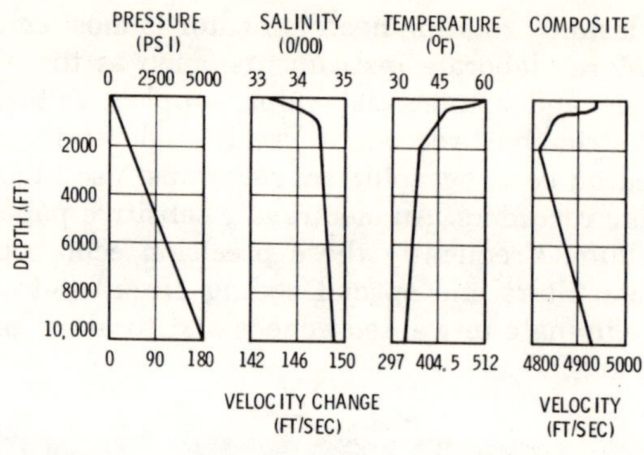

Courtesy U.S. Navy

Fig. 1-7. Pressure, salinity, and temperature as a function of depth.

The echo sounder is an important tool for the biological oceanographer and commercial fisherman. Echo sounders with high resolution are used to find and track schools of fish.

THE BATHYTHERMOGRAPH

Another landmark in the development of oceanographic instrumentation was the invention of the bathythermograph by Athelstean Spilhaus in 1938. Thermal conditions are of great importance in determining the layer depth and the velocity of propagation of sound in the sea. While the speed of sound in seawater is affected by pressure, salinity, and temperature, temperature has the greatest effect. The speed of sound through water increases between four and eight feet per second for each degree Fahrenheit *decrease* in temperature. Sound waves will travel toward the cooler layers of water. Since the temperature profile of the sea normally decreases with depth, the normal path of sound is in a downward direction.

This continues through the isothermal region (Fig. 1-8) to a zone where the temperature decreases rapidly with depth. This zone is called the thermocline. Past the thermocline, the decrease of temperature with depth is gradual.

Since the thermocline area has great effect on sound transmission, it is important to locate it accurately. Sound originating above the thermocline is reflected back toward the surface, while sound below the thermocline is frequently undetectable. In war, submarines try to locate the thermocline and remain below it to prevent being detected by searching surface vessels. It is equally important to researchers who are studying the propagation of sound in the ocean or are tracking marine animals.

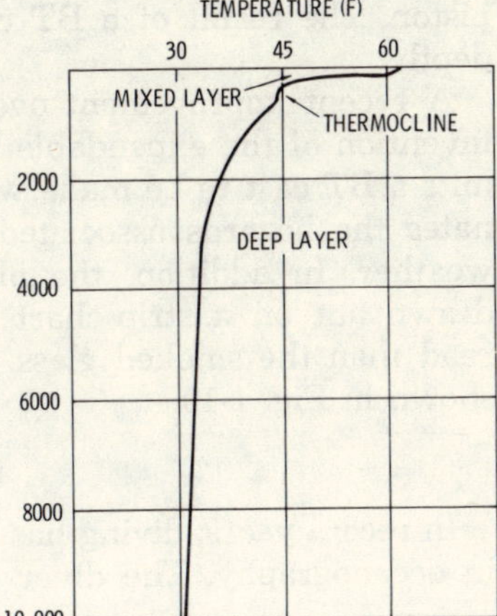

Fig. 1-8. Ocean thermal layers.

The bathythermograph, or *BT* as it is popularly known, consists of a temperature element and a pressure sensor (Fig. 1-9). The temperature element is made up of a fine copper tube filled with xylene. The copper tube is wound around the tail section where it is in constant contact with the seawater. Inside the unit is a Bourdon tube and a pressure piston. As the xylene expands or contracts due to the temperature changes, the tubing expands or contracts, causing the arm attached to the Bourdan tube to move.

The stylus records the effects of the temperature changes on a small smoked-glass slide. The slide is moved by the pressure

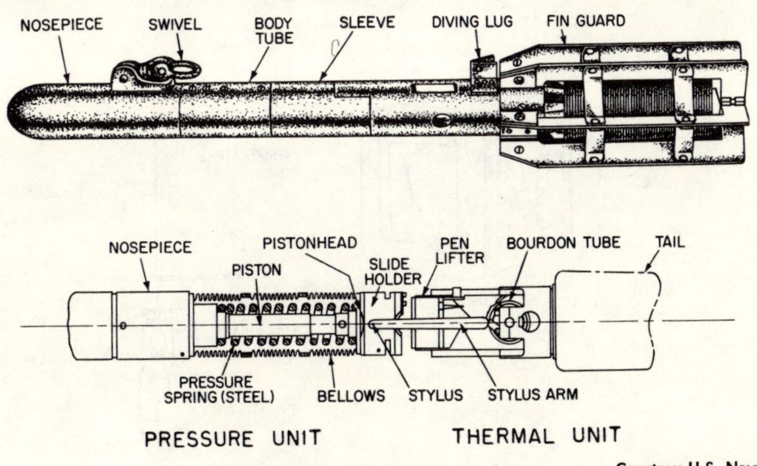

Fig. 1-9. The bathythermograph.

piston. The result of a BT cast is a plot of temperature versus depth.

A recent improvement over the mechanical BT has been the invention of the expendable bathythermograph. This device permits a BT cast to be made without slowing the vessel and eliminates the hazards associated with recovering the BT in rough weather. In addition, the plot of temperature versus depth is drawn out on a strip-chart recorder, which is much easier to read than the smoked glass slide. An expendable BT system is shown in Fig. 1-10.

DIVING

In recent years, diving has begun to play a very important role in oceanography. The diver fills a need that cannot be met by

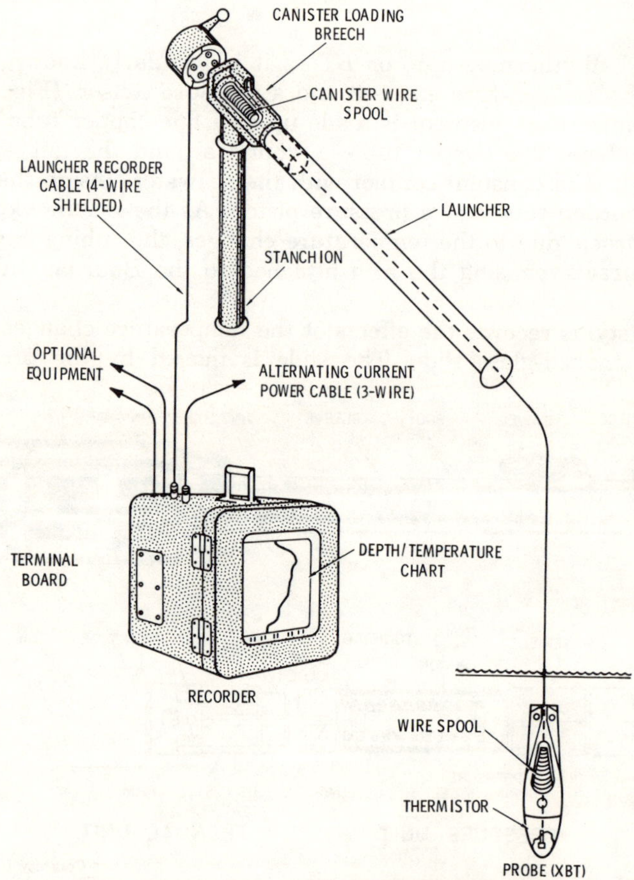

Fig. 1-10. The expendable bathythermograph, "XBT."

either scientific instruments or research submersibles. Diving has been practiced since ancient times. Even Greek mythology abounds in legends involving diving. Xerxes used divers in naval warfare. Many everyday items in the Greek and Roman empires, such as sponges and the pigment for the royal purple dye (obtained from a shellfish), were obtained by diving.

Early diving was concerned with getting items for use or trade, or with recovery of material from sunken vessels. It was natural that those involved in diving would recount their experiences, thus adding to the folklore and the knowledge of the ocean depths.

The industrial revolution saw great improvements in diving apparatus. The steam-driven compressor and articulated diving suit were examples of nineteenth-century developments. Unfortunately, the cost of diving operations and the cumbersome equipment made it impractical for scientific research of that day.

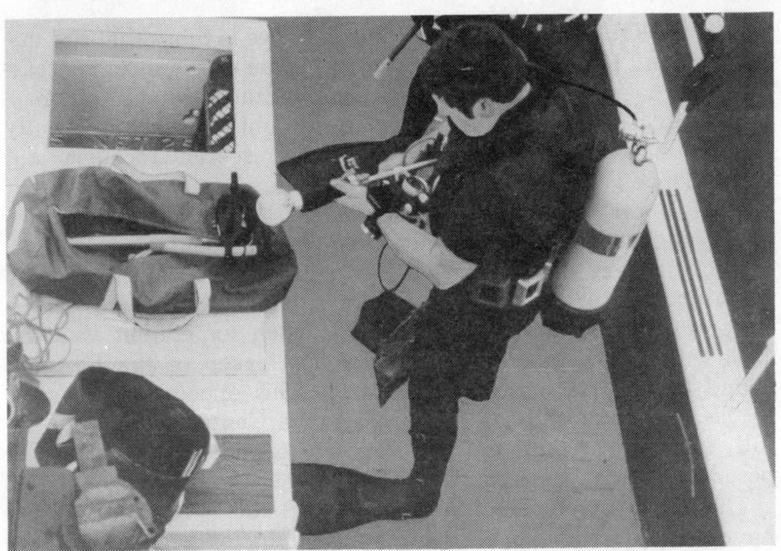

Fig. 1-11. A research diver preparing his underwater camera.

It was not until well into the twentieth century that the development of self-contained, low-cost diving equipment reached the practical stage. Accessories such as faceplates and fins were developed to a high degree. Successful goggles were mass manufactured in the 1930s. These soon evolved into a single-viewing plane type. While this eliminated the double vision problem of the goggles, the diver still suffered from pressure problems as he descended. Many solutions were tried, such as rubber bulbs, to prevent the faceplate from being painfully forced into the face. When the solution was found, it was incredibly simple—the mask was simply made large enough to cover the nose!

The major breakthrough came with the development of self-contained, underwater breathing apparatus, *SCUBA*. The French led the world in this development. One of the first units to appear was designed by Commander Yves Le Prier of the French navy. It was simply a compressed air bottle connected to a face mask. The diver manually operated a valve when he wanted to breathe. The best design was the result of work done by Emile Gagnan and Captain Jacque Yves Cousteau. The design is the now famous, fully automatic valve that meters compressed air on demand. This invention has been properly nicknamed "the passport to inner space" (Fig. 1-11).

UNMANNED REMOTE-CONTROLLED UNDERWATER VEHICLES

A remarkable class of underwater vehicles is those that are unmanned and tethered. Of this category, some of the most successful have been developed by the United States Naval Undersea Center in San Diego, California. These vehicles were originally conceived for use primarily as underwater search, recovery and surveillance vehicles. The best known of this family is the *CURV*. CURV is the acronym for *C*able Controlled *U*nderwater *R*ecovery *V*ehicle. This vehicle achieved fame on several occasions: in the recovery of a hydrogen bomb lost off Palomares, Spain; in the rescue of the British midget submarine Pisces; and for support of scientific expeditions. One such expedition was the 1974 Interstate Electronic Corporation's survey of the Farallon Island radioactive waste disposal area. This expedition, directed by Larry Brady and his Naval Undersea Center team, successfully located in 850 meters of water, cans of radioactive waste disposed of twenty years earlier!

The CURV vehicles illustrate remarkable versatility. They have a major advantage over manned vehicles because no life-support system is required and they can be operated for extensive periods of time without having to surface. They are more economical to operate than a manned submersible, and there is no risk to life.

In some ways, the CURV is like an upside-down helicopter (Fig. 1-12). The vehicle itself is slightly positive in buoyancy. The first sections of cable and line are equipped with floats to make them positively buoyant. In operation, one of the three 10-horsepower motors pulls the vehicle down toward the ocean floor. Positioning and control are accomplished by the other two

Fig. 1-12. CURV III, the U.S. Navy's cable-controlled underwater recovery vehicle.

motors. The nominal operation depth is 7000 feet, with an emergency capability of 10,000 feet.

The CURV is a highly complex system. It contains a sophisticated high-resolution sonar system, two closed-circuit television cameras, a 35-mm underwater camera, and special high-intensity underwater lighting.

A highly sophisticated hydraulically operated manipulator is used for recovering objects, manipulating tools, and obtaining sediment samples. Since the operator is able to observe the operation of the manipulator, sediment samples can be taken with great care and the upper few millimeters of sediment can be preserved. This type of sample is required for accurate measurement of man-made radioactivity.

The operator manipulates the controls from a self-contained control van aboard the support vessel (Fig. 1-13). He can accurately maneuver the vehicle while observing the target on a sonar display and two television monitors. Other instruments display the vehicle depth, the remaining distance to the bottom, the water temperature, and the status of the propulsion system.

The *CURV III* system is air mobile and readily transportable. It is entirely self-contained and can be operated from a wide variety of vessels, such as the *M/V Gear* (Fig. 1-14) supplied by the U.S. Navy's Supervisor of Salvage for the Farallon Island

Courtesy U.S. Naval Undersea Center

Fig. 1-13. CURV III support ship, XFNX-30.

Fig. 1-14. The motor vessel M/V Gear lying to in San Francisco Bay.

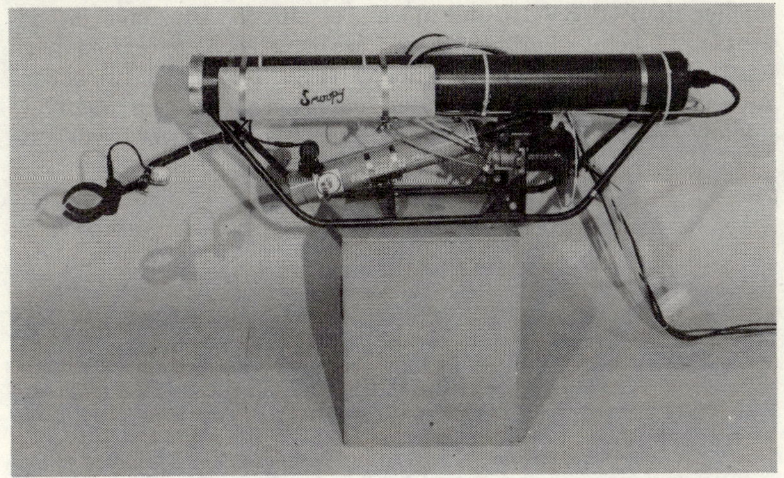

Courtesy U.S. Naval Undersea Center

Fig. 1-15. "Snoopy," a remote-controlled underwater vehicle.

expedition. Ships like this can readily accommodate the CURVE III on short notice.

Other members of the Naval Undersea Center's family of tethered vehicles are Snoopy and SCAT. Both of these are much

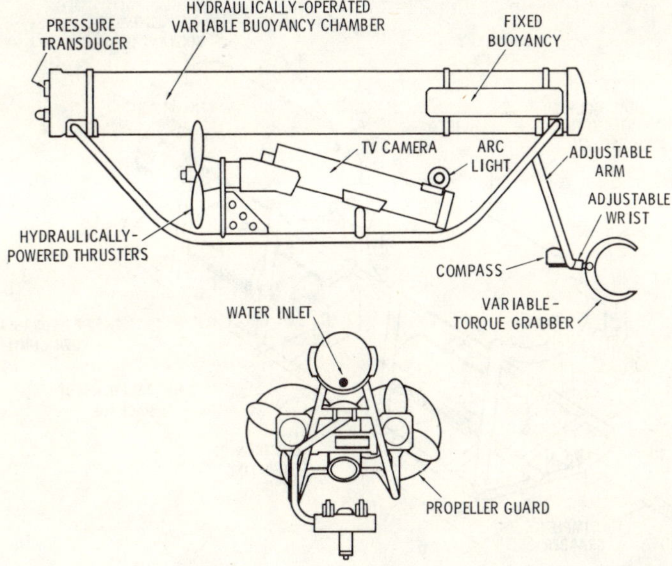

Courtesy U.S. Naval Undersea Center

Fig. 1-16. Snoopy equipment layout.

smaller than CURV III and more specialized. They have the advantage of being lightweight and are capable of operation from practically any small vessel. Snoopy (Figs. 1-15 and 1-16) is configured primarily for underwater television viewing in relatively shallow water. It is powered by two ⅕-horsepower hydraulic

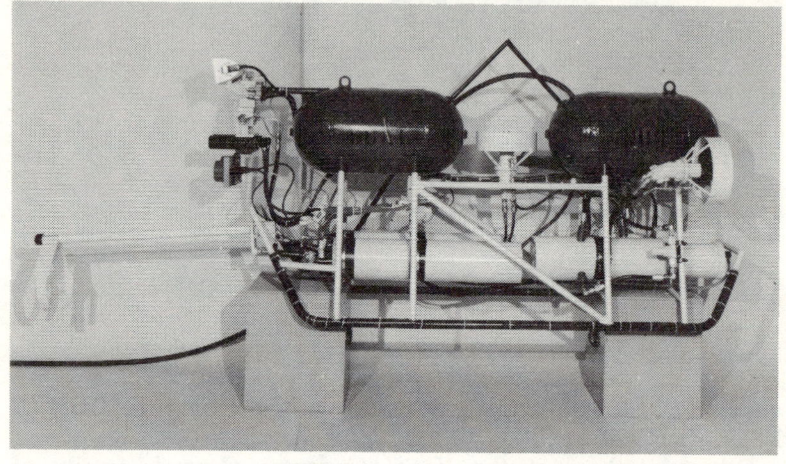

Courtesy U.S. Naval Undersea Center

Fig. 1-17. The U.S. Navy's remote-controlled underwater vehicle SCAT.

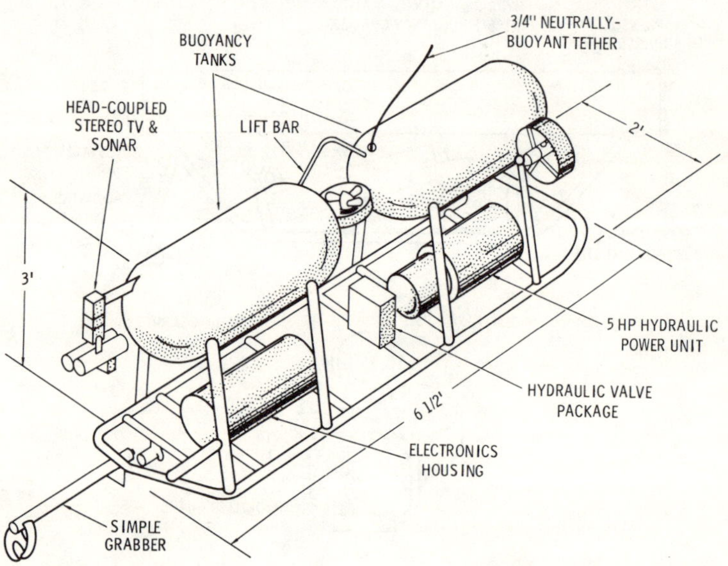

Courtesy U.S. Naval Undersea Center

Fig. 1-18. SCAT equipment layout.

motors for horizontal movement and has a hydraulically operated variable-buoyancy chamber for vertical movement. Its normal speed is 1.5 knots, with an operating depth of 100 feet.

Snoopy's hydraulic thrusters are independently controllable. Once set, the depth controls are designed to automatically maintain a given depth. Snoopy's 50-pound weight is easily handled by one man.

SCAT (Fig. 1-17) is larger than Snoopy, with capabilities between Snoopy and CURV. The acronym stands for *S*ubmersible *C*able *A*ctuated *T*eleoperator. It is hydraulically powered, with power being supplied by a 5-horsepower, 440-V ac-driven fixed-

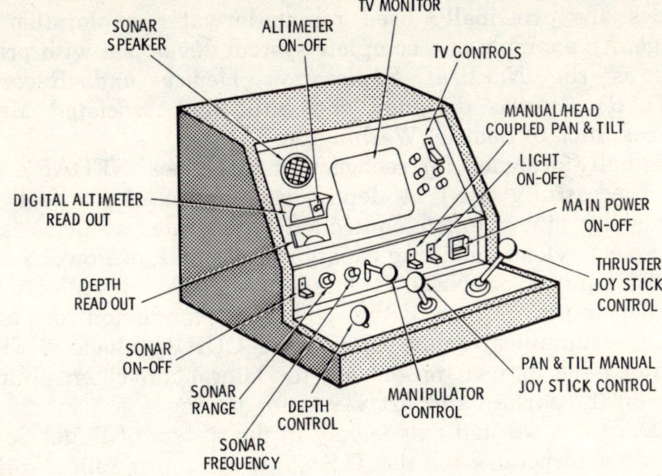

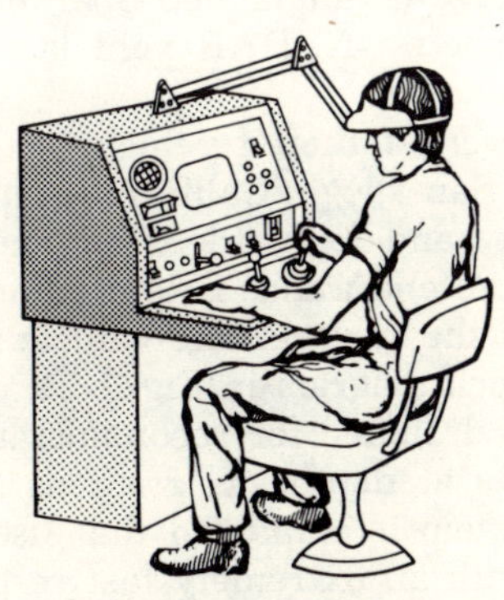

Courtesy U.S. Naval Undersea Center

Fig. 1-19. SCAT control console.

displacement hydraulic pump. Its rated operating depth is 2000 feet. In addition to a search sonar, the system incorporates dual television cameras, configured to permit stereoscope viewing, and a hydraulic manipulator. (See Figs. 1-18 and 1-19.)

The controls are located in a sophisticated console where the operator can observe undersea progress on two miniature cathode-ray tubes mounted in a special lightweight headset. The cameras automatically follow the movement of the operator's head! This results in a unique interactive manipulative display with a quick response capability.

Commercial firms have also developed remote-controlled tethered vehicles. Like the government-developed systems, these vehicles are principally used for underwater exploration and salvage. An example of a complete system developed with private funds is the *Nautical Exploration Device and Recoverer* (NEDAR). It was designed and built by Associated Marine Services, Inc., of Seattle, Washington.

Originally designed for recovery of torpedoes, NEDAR's have been successfully used at depths ranging to 18,000 feet. Unmatched by other remote-controlled underwater vehicles is the NEDAR's payload, or lifting capability, of one ton. Power for the hydraulic pump on NEDAR I is provided by a 440-volt submersible motor. Motor drive power is conducted on a 16-conductor umbilical cable. Unlike the CURV vehicle, NEDAR does not have its own propulsion for lateral movement, but depends on the surface support vessel for towing.

NEDAR has worked extensively in the waters of Puget Sound, recovering torpedoes for the U.S. Navy test program. Working in deep water, NEDAR has investigated deep-water mineral deposits. The advantage of unmanned operation was demonstrated during one dive when a NEDAR vehicle operated continuously at 2300 feet for 56 hours.

Another commercial tethered vehicle is the RCV-125 built by Hydroproducts of San Diego, California. This vehicle is designed for high portability and for work is confined areas. As with all tethered vehicles, its endurance is limited only by the endurance of the operators at the surface. The vehicle is shown in Fig. 1-20, and the system block diagram in Fig. 1-21.

The RCV-125 is designed for reconnaissance and inspection of underwater objects at depths as great as 2000 meters! A low-light television system is employed that uses a very clever fish-eye lens system with an extremely fast response time. The view angle can be pitched ±90 degrees from the horizon. Two 45-watt tungsten halogen lamps provide sufficient illumination for a viewing range of 7 to 10 meters. A carefully thought out feature is a reflector for the lamps that can be adjusted to minimize the scattering effect encountered in turbid water.

Maneuverability of the vehicle is outstanding. Four electric motors provide three degrees of freedom in translation (forward and reverse, left and right, up and down) and one rotational degree of freedom (heading rotation).

Courtesy Hydro Products

Fig. 1-20. The RCV-125, a commercial tethered vehicle.

The control station provides the operator with a tv monitor and a joy-stick type of control. Once set, the servo system maintains the depth and heading. Vehicle-motion parameters, such as depth, heading, and lens-pitch angle, are continuously displayed on the console and simultaneously recorded by a video tape recorder.

MANNED SUBMERSIBLES

The 1960s can be described as the golden age of research submersibles. A side benefit of the space program was an increased awareness of the lack of knowledge about hydrospace. Many aerospace companies projected future business in the field of

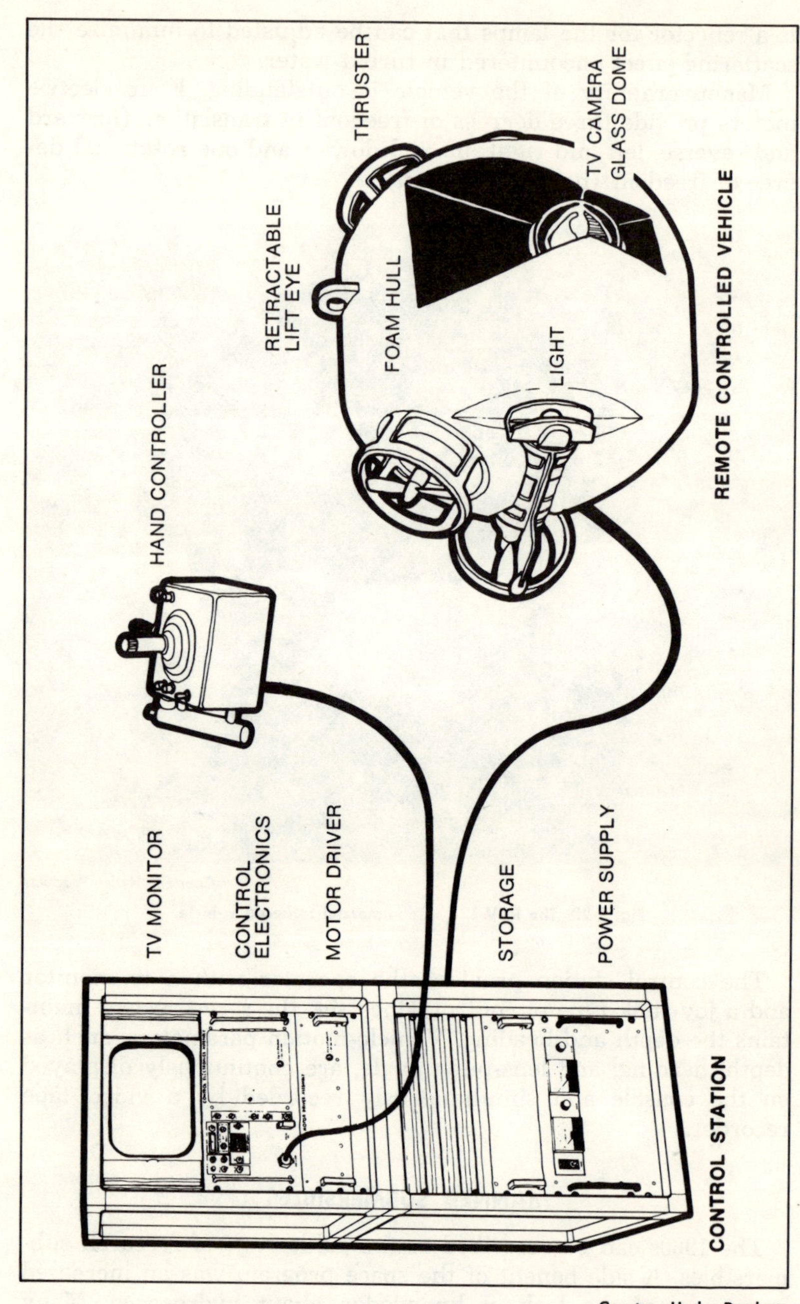

Courtesy Hydro Products

Fig. 1-21. System diagram, RCV-125.

oceanic exploration and invested heavily in the development and construction of submarines.

By the end of the decade, the dreams were shattered and most of the vessels had been either mothballed or cannibalized for other programs. Those vessels showing the greatest survivability were small ones that could be used economically for commercial work such as inspection of sewer outfalls, recovery of lost objects, and shallow-water salvage jobs.

One of the more sophisticated research submersibles in the Deep Quest. Designed and built by Lockheed Missiles and Space Company as an engineering and oceanographic research submersible, it has been fully certified by the U.S. Navy for operation at 8000 feet. The Deep Quest and its support ship, the TransQuest (Fig. 1-22), combine to provide an advanced research system.

The vessel is quite spacious. It will accommodate two or three scientific observers in addition to the pilot and navigator. The specially designed life support and environmental control systems provide comfortable submerged operation for up to 48 hours. An emergency system with closed-loop facemasks provides up to three hours of survival time in a contaminated atmosphere.

The vessel can maintain a submerged speed of four knots. Precise positioning and depth control are accomplished with fore

Courtesy Lockheed Missiles and Space Co.

Fig. 1-22. The Deep Quest on board its support ship, TransQuest.

and aft vertical thrusters, lateral water jets, and a shot ballast system controlled by an integrated control system. Attitude and trim can be controlled within the limits of ±30° in trim and ±10° of list by transferring liquid mercury between tanks.

Courtesy Southwest Research Institute

Fig. 1-23. The NEMO submersible.

The observation and sampling equipment are also quite sophisticated. A continuous transmission frequency-modulated (ctfm) sonar system of advanced design is mounted on the control panel to assist the pilot and navigator in searching for underwater objects and in avoiding collisions. Two pan-and-tilt television cameras provide the crew with a view of the near bottom terrain. Fixed photography of high quality is possible with the high-intensity external lights.

Courtesy Southwest Research Institute

Fig. 1-24. The Deepview submersible.

Accessories include a piston corer for taking sediment samples, a vane shear device capable of monitoring sediment shear strength, manipulators, and equipment for measuring the chemical and acoustical properties of the water mass.

The operational cost of a manned submersible is high, and there is no way of completely eliminating the risk to human life. Still, there are applications that can only be met by having a skilled human operator on the bottom.

Other manned submersibles still in operation are the NEMO (Fig. 1-23) and Deepview (Fig. 1-24). These vessels are operated by the Ocean Engineering Department of the Southwest Research Institute of San Antonio, Texas.

The NEMO is a strange-looking vehicle consisting of an acrylic hull mounted on an integral winch and anchor package. The acronym stands for *N*aval *E*xperimental *M*anned *O*bservatory.

Basically, NEMO is a two-man, untethered submersible with both horizontal and vertical maneuverability. It is rated for depths to 600 feet. An acrylic pressure hull is supported by a structural cage mounted on a pod containing the ballast tank service module and main battery pack.

The sphere is 66 inches in diameter and is 2 1/2 inches thick. The vehicle can ascend and descend at rates up to 60 feet/second; its lateral speed is 2 knots. The life support system is designed for 6 hours of operation and 48 hours of survivability.

Deepview (Fig. 1-24) is a more conventional-looking manned submersible designed to operate in depths to 2000 feet. The hull is a ring-reinforced, steel pressure hull with a four-inch-thick, 48-inch-diameter Plexiglas hemisphere at the bow. Maximum speed is 3 knots, with a primary operations time of 24 hours and a survival time of 72 hours.

Another vehicle in the class of working small submersibles is the Nekton series. This series is composed of the Alpha, Beta, and Gamma. This class of submersibles was designed, built, and operated by General Oceanographics of Newport Beach, California. The Nektons are commercial work submarines rated for 1000-foot operational depths. Nominal underwater cruising speed is 2 knots, with a "flank" speed of 3.5 knots. Equipped with high-intensity light, sonar, and a manipulator arm, they are designed strictly as commercial submersibles for reconnaissance and operations. Reconnaissance tasks include geological mapping, sediment sampling, biological investigation, and salvage search. Nektons have been used for a wide variety of commercial jobs on ocean outfalls, wellheads, subsea pipelines, and pollution monitoring.

Texas A&M University's Department of Oceanography operates the PC-14C submersible "Diaphus" (Fig. 1-25). This boat

Courtesy Texas A&M University

Fig. 1-25. The PC-14C submersible.

Courtesy Texas A&M University

Fig. 1-26. Research vessel R/V Gyre.

Courtesy Submarine Development Group One, U.S. Navy

Fig. 1-27. Turtle, a manned submersible.

was built by Perry Submarine Builders. It carries a crew of two and operates at a depth of 1200 feet.

It is operated from the research vessel R/V Gyre (Fig. 1-26), but it is small enough to operate from a variety of ships of opportunity.

The Turtle (Fig. 1-27) and its sister craft, Sea Cliff, are deep diving research submersibles under the command of the U.S. Navy's Submarine Development Group 1. They have a depth capability of 7000 feet. They were designed for undersea exploration, search, and recovery. Powered by lead-acid batteries, these vessels can operate for ten hours at a speed of 1.5 knots.

2

Basic Concepts

TOPOGRAPHY

What little man has discovered about the topography of the ocean basins supports the belief that they are similar in structure to formations found on land. There are mountains, plains, canyons, and other familiar land forms. However, the general topography is much more rugged than that which is familiar to us on land. There are features quite unique to the oceans. In describing the topography of the ocean bed, the term *bathymetry* (bə-ˈthim-ə-trē) is used.

The oceanic zones are shown in Fig. 2-1. The continental shelf is the first of three major subsea bathymetric features. It is much like the rolling coastal plains that border many ocean areas. It has become of prime importance in recent years due to the increased demand for food and minerals from the sea. Sunlight penetrates to most of its depth, resulting in a high level of plant life and an abundance of food for marine animals. Many of the larger fisheries are located here.

It would be difficult to precisely define the limits of the continental shelf. Commonly it is defined as the zone from the mean high tide line to the 100-fathom contour. This definition results in a shelf that ranges in width from 0 to 800 miles, with an average width of 30 miles. The average slope angle is about two fathoms per mile, a slope that would not be perceptible to the human eye.

Moving seaward, the shelf becomes a slope, rapidly dropping to the ocean floor. This slope is frequently laced with submarine canyons of great size. These canyons were formed in prehistoric times by rivers.

Some recently identified phenomena of these canyons are large movements of mud and sand which erode the canyons even deeper. At the bottom of the continental slope lies the final benthic feature, the abyssal plain.

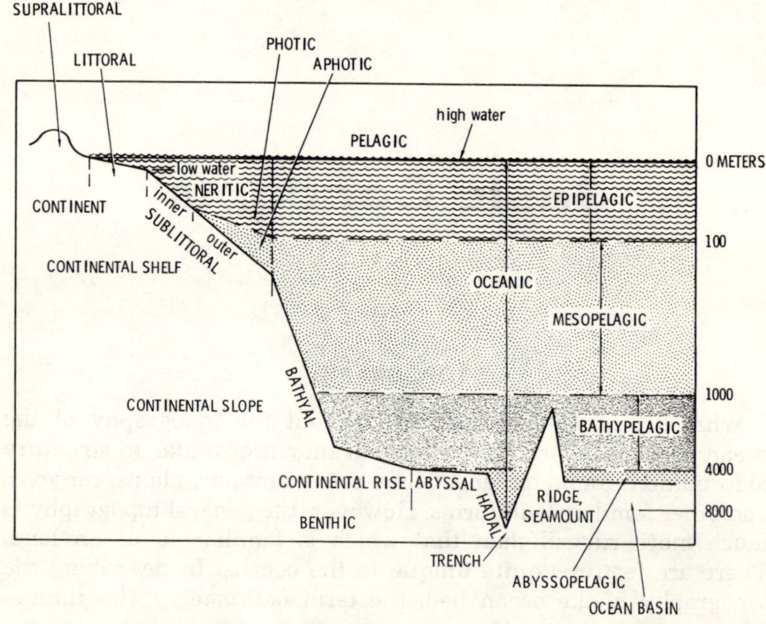

Fig. 2-1. Marine zones.

The more prominent terrain features include: volcanoes, guyots (flat-topped sea mounts named after the French geographer Arnold Guyot), atolls, sea mounts, submarine ridges, and canyons. A spectacular presentation of these features can be found on the conceptual bathymetric charts prepared by the National Geographic Society of the United States.

Guyots are flat-topped mountains whose level tops are found at depths as great as a half mile below the surface. They are of special interest to scientists since their flat tops exhibit marks indicating flattening by erosion. Core samples from the tops reveal interesting fossils of prehistoric land animals.

The oceanic environment is nominally divided into a series of zones. The first major subdivision is between aquatic (*pelagic*) and bottom (*benthic*) zones.

The dry land above the high water mark is named the *supralittoral* zone. Going seaward, the region between the high and low water marks is the *littoral* zone. This is a relatively shallow zone in which light can readily penetrate. Typically the maximum depth of the littoral zone of the water is in the order of ten meters (see Fig. 2-1).

Next is the *sublittoral* zone in which light starts to fade out and the depth increases gradually as the ground slopes toward the end of the continental shelf. The maximum water depth in this zone is generally assumed to be 200 meters.

From the sublittoral zone, the ground falls off rapidly to a depth of 3000 meters, to the final benthic division, the *abyssal* deep—the zone from 3000 meters to the deepest points of the ocean.

The aquatic portion, or *pelagic* region, is also divided into zones. The first is the *neritic*. This is the zone between the low water mark and approximately 200 meters. This includes the region in which light penetrates, the photic zone. Photosynthesis takes place here and the nuturing of the phytoplankton, the basic step in the food chain of oceanic animals. Most photosynthesis takes place in the upper 80 meters, an area termed the "true" photic or euphotic zone.

Next is the *epipelagic,* seaward from the neritic, and the uppermost oceanic environmental zone. It ranges from the surface to a depth of 100 meters. Marine life is concentrated in this zone. The *mesopelagic* zone is immediately below the *epipelagic* and ranges from a depth of 100 meters to 1000 meters. It is generally devoid of marine life at its lower boundary. The two final zones are the *bathypelagic,* 1000 to 4000 meters, and the *abyssopelagic,* greater than 4000 meters.

WAVES, TIDES AND CURRENTS

Waves are the most widely observed and exciting of ocean phenomena. They are generated in a variety of ways. Most are produced by the action of the wind. Other forces that generate waves are the gravitational forces of the sun, moon, and planets, and seismic disturbances. The former are familiarly associated with the tides, while the seismic disturbances are associated with the *Tsunami* or, as it is often called, the tidal wave.

In the study of waves, a specific terminology is used. Fig. 2-2 illustrates the most commonly encountered terms. Waves are most often classified by their period, the length of time required for one wavelength to pass a fixed point. The common classifications are listed in Table 2-1.

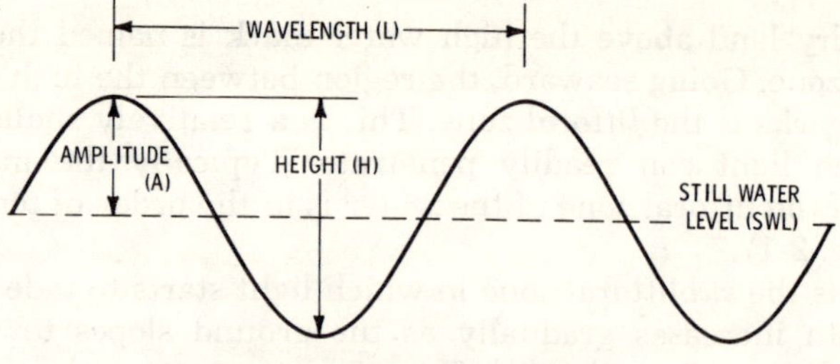

Fig. 2-2. Common wave terms.

The speed of a wave can be readily calculated if the wavelength and period are known. Simply divide the wavelength by the period as shown in the following formula:

$$C = \frac{L}{T}$$

where,

C is the speed of the wave,
L is the length of the wave,
T is the period of the wave.

From the classifications shown in Table 2-1, it is evident that waves come in a wide range of sizes. They also come in an endless variety of forms. Because of their randomness and complexity, it is not possible to develop a simple, accurate description for analytical purposes. In order to have a basis for discussion, oceanographers frequently use a simplified or ideal wave such as the one shown in Fig. 2-2. For small waves with rounded crests, the mathematical expressions for a sine wave are used. For the steeper and more commonly encountered waves, the mathematical form known as a trochoid is used.

Since most waves are generated from the action of the winds, a skilled observer can often judge the wind speed simply by looking at the waves. Table 2-2 defines the relationship between the wind

Table 2-1. Classification of Waves by Period

Classification	Period
Capillary	less than 0.1 second
Ultragravity	0.1 second to 1 second
Gravity	1 second to 30 seconds
Infragravity	30 seconds to 5 minutes
Long period	5 minutes to 12 hours
Tidal	12 hours to 24 hours
Transtidal	greater than 24 hours

speed and the observable effects. The Beaufort number shown to the left of the table listings was developed by Sir Francis Beaufort of the Royal Navy in 1806 as an aid in determining the maximum amount of sail practical in a given wind. The numbering system is in wide use today as an index for reporting sea conditions.

The size of wind-generated waves depends upon the wind speed, the maximum distance the wind blows in a constant direction (called the fetch), and the length of time it blows across the fetch. This relationship is shown in Fig. 2-3. From the figure it would seem to be a simple task to accurately predict wave height. Unfortunately such predictions are risky and complex, since the wind is seldom steady in speed or direction. In addition, there are limits to the size of the wave. After a point, the wave becomes so steep that it becomes unstable and will break. This point commonly occurs when the angle between the sides of the crest is less than 120 degrees or when the ratio of wave height to wave length exceeds 1: 7.

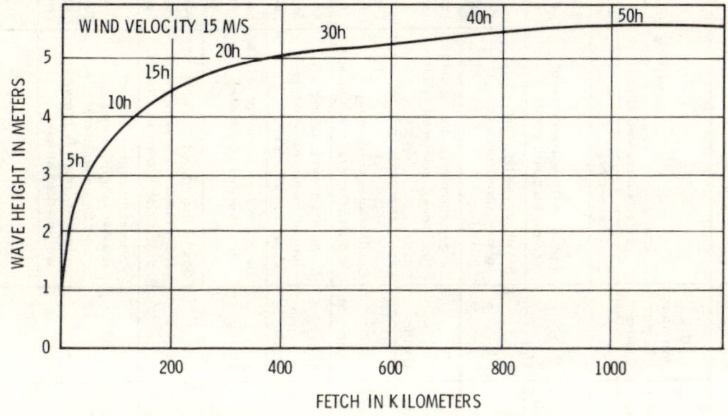

Source: The Oceans, Sverdrup, et al.

Fig. 2-3. Growth of waves with fetch and duration.

Tides may be classified as long-period waves caused by the forces of the moon, sun, or other celestial bodies. The moon is the main causing body, but at certain locations (primarily the south Pacific) the solar tide may become more important. The periodic rise and fall of water associated with the tidal phenomenon occurs, twice daily at most places. It exhibits a wide range of variations during a lunar month. Examples of these variations are given in Fig. 2-4. When the moon and sun are in line and pulling together (Fig. 2-5A), maximum tides occur. These are called spring tides. When the forces of the moon and sun are in quadrature (Fig. 2-5B), the smaller neap tides occur.

Table 2-2. Beaufort Scale With Corresponding Sea State Codes

Wind speed						U.S. Weather Bureau term	Estimating wind speed		Hydrographic Office		International	
Beaufort number	knots	mph	meters per second	km per hour	Seaman's term		Effects observed at sea	Effects observed on land	Term and height of waves, in feet	Code	Term and height of waves, in feet	Code
0	under 1	under 1	0.0-0.2	under 1	Calm	Light	Sea like mirror.	Calm; smoke rises vertically.	Calm, 0	0	Calm, glassy, 0	0
1	1-3	1-3	0.3-1.5	1-5	Light air		Ripples with appearance of scales; no foam crests.	Smoke drift indicates wind direction; vanes do not move.	Smooth, less than 1	1		
2	4-6	4-7	1.6-3.3	6-11	Light breeze		Small wavelets; crests of glassy appearance, not breaking.	Wind felt on face; leaves rustle; vanes begin to move.	Slight, 1-3	2	Rippled, 0-1	1
3	7-10	8-12	3.4-5.4	12-19	Gentle breeze	Gentle	Large wavelets; crests begin to break; scattered whitecaps.	Leaves, small twigs in constant motion; light flags extended.	Moderate, 3-5	3	Smooth, 1-2	2
4	11-16	13-18	5.5-7.9	20-28	Moderate breeze	Moderate	Small waves, becoming longer; numerous whitecaps.	Dust, leaves, and loose paper raised up; small branches move.	Rough, 5-8	4	Slight, 2-4	3
5	17-21	19-24	8.0-10.7	29-38	Fresh breeze	Fresh	Moderate waves, taking longer form; many whitecaps; some spray.	Small trees in leaf begin to sway.			Moderate, 4-8	4
6	22-27	25-31	10.8-13.8	39-49	Strong breeze	Strong	Larger waves forming; whitecaps everywhere; more spray.	Larger branches of trees in motion; whistling heard in wires.			Rough, 8-13	5
7	28-33	32-38	13.9-17.1	50-61	Moderate gale		Sea heaps up; white foam from breaking waves begins to be blown in streaks.	Whole trees in motion; resistance felt in walking against wind.	Very rough, 8-12	5	Very rough, 13-20	6
8	34-40	39-46	17.2-20.7	62-74	Fresh gale	Gale	Moderately high waves of greater length; edges of crests begin to break into spindrift; foam is blown in well-marked streaks.	Twigs and small branches broken off trees; progress generally impeded.				

9	41-47	47-54	20.8-24.4	75-88	Strong gale		High waves; sea begins to roll; dense streaks of foam; spray may reduce visibility.	Slight structural damage occurs; slate blown from roofs.	High, 12-20			6
10	48-55	55-63	24.5-28.4	89-102	Whole gale	Whole gale	Very high waves with overhanging crests; sea takes white appearance as foam is blown in very dense streaks; rolling is heavy and visibility reduced.	Seldom experienced on land; trees broken or uprooted; considerable structural damage occurs.	Very high, 20-40	7	High, 20-30	7
11	56-63	64-72	28.5-32.6	103-117	Storm		Exceptionally high waves; sea covered with white foam patches; visibility still more reduced.		Mountainous, 40 and higher	8	Very high, 30-45	8
12 13 14 15 16 17	64-71 72-80 81-89 90-99 100-108 109-118	73-82 83-92 93-103 104-114 115-125 126-136	32.7-36.9 37.0-41.4 41.5-46.1 46.2-50.9 51.0-56.0 56.1-61.2	118-133 134-149 150-166 167-183 184-201 202-220	Hurricane	Hurricane	Air filled with foam; sea completely white with driving spray; visibility greatly reduced.	Very rarely experienced on land; usually accompanied by widespread damage.	Confused	9	Phenomenal, over 45	9

Note: Since January 1, 1955, weather map symbols have been based upon wind speed in knots, at five-knot intervals, rather than upon Beaufort number.

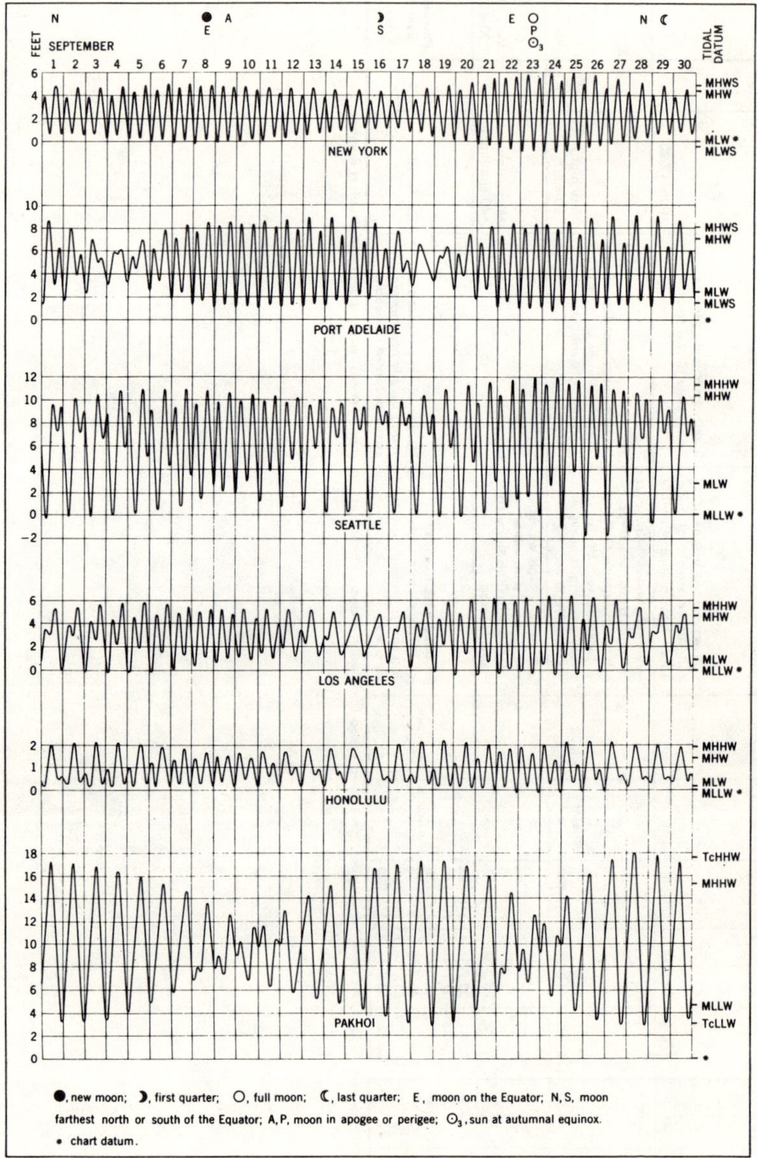

Fig. 2-4. Tidal variations.

Because the tides greatly influence commerce, they have been studied since the earliest times. Early measurements simply consisted of erecting a pole, marked at set intervals, in the harbor. Timely observations were made and recorded. From this information, predictions were made to assist mariners. More recently, electromechanical automatic recording tide gauges have been used that recorded with a pen on a paper-covered drum. More modern instruments are entirely electronic and enter the information directly into electronic computers.

Ocean currents are water movements in a predominately horizontal direction. Like waves, ocean currents are generated mainly by the winds. However, another major causing force is the differences in density encountered in various parts of the ocean.

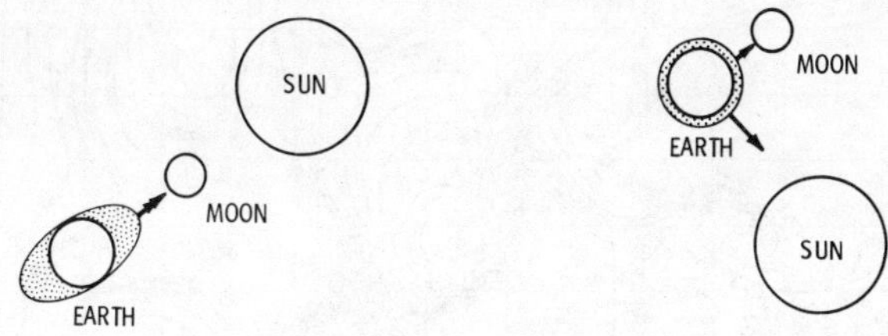

Fig. 2-5. Tides and celestial bodies.

Probably the best known of the ocean currents is the Gulf Stream, the "river in the ocean" that Lt. Maury described in 1858 in his book *The Physical Geography of the Sea.* A number of ocean currents flow with extreme regularity, setting up predictable circulation patterns. The more important of these currents are shown in Fig. 2-6. Generation of a current by force of the wind requires a constant wind blowing for a long period. Therefore, many of the major currents are directly associated with prevailing winds of great duration. The trade winds, for example, generate the equatorial currents which span as much at 50 degrees of latitude. Interestingly enough, the current generated by the force of the wind does not flow in the direction of the wind! The coriolis force deflects the current to the right in the Northern Hemisphere and toward the left in the Southern Hemisphere.

Many of the ocean currents affect the climate of the coasts that they flow along. The warm water of the Gulf Stream warms the southwest coast of Iceland so much that Reykjavik has a higher average winter temperature than New York City, which is far to the south.

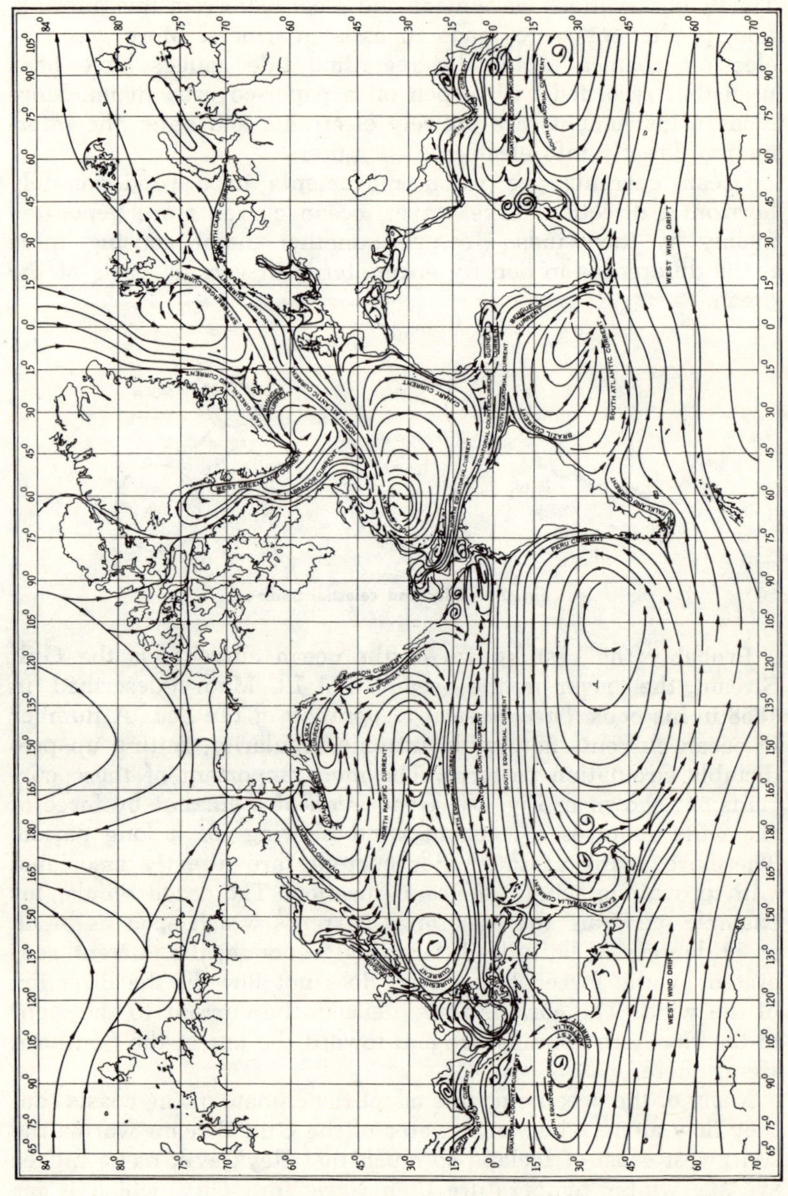

Fig. 2-6. Major world surface currents.

3

Units of Measurements

Before we can measure any quantity, we must have a scale on which to measure it. For example, if we wanted to measure a length of cloth, we would commonly express it in terms of a scale divided into an equal number of parts, such as a yardstick divided into feet and inches. Every craft from carpentry to physics has, from time to time, adopted units of measure convenient for its own purposes. In many cases, the basic units of measurement varied widely from country to country. As science and technology evolved, weights and measures became more complex, and the variation in standards between countries and crafts caused many problems when it came to exchanging information.

The English measurement system, commonly used in the United States, is one of the most difficult to adapt to scientific purposes. The need for a single international coordinated measurement system was recognized back in the 1600s. In 1790, the French Academy of Sciences chartered a commission to develop a standard of measurements that could be truly international. The system was extremely simple and scientific. The unit of length was related to a proportion of the earth's circumference; units of volume and mass were derived from the unit of length. Therefore, all the basic units were related to each other and to nature. Other units were created by multiplying or dividing the basic units by 10 or multiples of 10.

The French commission assigned the name "metre" to the unit of length. They derived this name from the Greek word *metron* meaning a measure. The metre was constructed so that it would equal one 10-millionth of the distance from the North Pole to the equator. The metric unit of mass, called the gram, was defined as a mass of one cubic centimeter of water at the temperature of maximum density. The liter was chosen as a unit of food capacity and was defined as a cube ten centimeters on a side.

Although the metric system spread rapidly through Europe after its compulsory adoption by France in 1840, and the United States adopted it by an Act of Congress in 1866, it has yet to be implemented in the United States. In 1971 Congress authorized a three year study on how to implement the system. We can anticipate that in a few years this system will be in widespread use throughout the United States.

The common conversion factors most frequently needed in oceanographic work are shown in Table 3-1.

Table 3-1. Conversions for Common Units

Unit	English		Metric	
fathom	6	feet	1.828	meters
cable	720	feet	219.45	meters
statute mile	5280	feet	1609	meters
nautical mile	6076	feet	1852	meters
foot	12	inches	0.3048	meter
yard	3	feet	0.9144	meter
meter	3.28	feet	100	centimeters
kilometer	3280	feet	1000	meters
pound	16	ounces	0.454	kilogram
gallon	4	quarts	3.785	liters
liter	1.05	quarts	1000	cubic centimeters

Just as the English system is used for weights and measures in the United States, the Fahrenheit scale is commonly encountered in temperature measurements. This scale does not lend itself well to scientific investigation. The most commonly used scale in science is the Celsius scale, named after the astronomer who developed the scale in 1742. Before the term *Celsius* was adopted in 1948 by international agreement, the scale was called centigrade. This name was derived from the fact that the scale between the freezing and boiling points of water was divided into 100 equal parts. Table 3-2 provides the conversion between the Farenheit and Celsius scales.

Table 3-2. Temperature conversion—Fahrenheit, Celsius and Kelvin Scales.

Conversion Tables for Thermometer Scales

F=Fahrenheit, C=Celsius (centigrade), K=Kelvin

F	C	K	F	C	K	C	F	K	K	F	C
°	°	°	°	°	°	°	°	°	°	°	°
−20	−28. 9	244. 3	+40	+4. 4	277. 6	−25	−13. 0	248. 2	250	−9. 7	−23. 2
19	28. 3	244. 8	41	5. 0	278. 2	24	11. 2	249. 2	251	7. 9	22. 2
18	27. 8	245. 4	42	5. 6	278. 7	23	9. 4	250. 2	252	6. 1	21. 2
17	27. 2	245. 9	43	6. 1	279. 3	22	7. 6	251. 2	253	4. 3	20. 2
16	26. 7	246. 5	44	6. 7	279. 8	21	5. 8	252. 2	254	2. 5	19. 2
−15	−26. 1	247. 0	+45	+7. 2	280. 4	−20	−4. 0	253. 2	255	−0. 7	−18. 2
14	25. 6	247. 6	46	7. 8	280. 9	19	2. 2	254. 2	256	+1. 1	17. 2
13	25. 0	248. 2	47	8. 3	281. 5	18	−0. 4	255. 2	257	2. 9	16. 2
12	24. 4	248. 7	48	8. 9	282. 0	17	+1. 4	256. 2	258	4. 7	15. 2
11	23. 9	249. 3	49	9. 4	282. 6	16	3. 2	257. 2	259	6. 5	14. 2
−10	−23. 3	249. 8	+50	+10. 0	283. 2	−15	+5. 0	258. 2	260	+8. 3	−13. 2
9	22. 8	250. 4	51	10. 6	283. 7	14	6. 8	259. 2	261	10. 1	12. 2
8	22. 2	250. 9	52	11. 1	284. 3	13	8. 6	260. 2	262	11. 9	11. 2
7	21. 7	251. 5	53	11. 7	284. 8	12	10. 4	261. 2	263	13. 7	10. 2
6	21. 1	252. 0	54	12. 2	285. 4	11	12. 2	262. 2	264	15. 5	9. 2
−5	−20. 6	252. 6	+55	+12. 8	285. 9	−10	+14. 0	263. 2	265	+17. 3	−8. 2
4	20. 0	253. 2	56	13. 3	286. 5	9	15. 8	264. 2	266	19. 1	7. 2
3	19. 4	253. 7	57	13. 9	287. 0	8	17. 6	265. 2	267	20. 9	6. 2
2	18. 9	254. 3	58	14. 4	287. 6	7	19. 4	266. 2	268	22. 7	5. 2
−1	18. 3	254. 8	59	15. 0	288. 2	6	21. 2	267. 2	269	24. 5	4. 2
0	−17. 8	255. 4	+60	+15. 6	288. 7	−5	+23. 0	268. 2	270	+26. 3	−3. 2
+1	17. 2	255. 9	61	16. 1	289. 3	4	24. 8	269. 2	271	28. 1	2. 2
2	16. 7	256. 5	62	16. 7	289. 8	3	26. 6	270. 2	272	29. 9	1. 2
3	16. 1	257. 0	63	17. 2	290. 4	2	28. 4	271. 2	273	31. 7	−0. 2
4	15. 6	257. 6	64	17. 8	290. 9	−1	30. 2	272. 2	274	33. 5	+0. 8
+5	−15. 0	258. 2	+65	+18. 3	291. 5	0	+32. 0	273. 2	275	+35. 3	+1. 8
6	14. 4	258. 7	66	18. 9	292. 0	+1	33. 8	274. 2	276	37. 1	2. 8
7	13. 9	259. 3	67	19. 4	292. 6	2	35. 6	275. 2	277	38. 9	3. 8
8	13. 3	259. 8	68	20. 0	293. 2	3	37. 4	276. 2	278	40. 7	4. 8
9	12. 8	260. 4	69	20. 6	293. 7	4	39. 2	277. 2	279	42. 5	5. 8
+10	−12. 2	260. 9	+70	+21. 1	294. 3	+5	+41. 0	278. 2	280	+44. 3	+6. 8
11	11. 7	261. 5	71	21. 7	294. 8	6	42. 8	279. 2	281	46. 1	7. 8
12	11. 1	262. 0	72	22. 2	295. 4	7	44. 6	280. 2	282	47. 9	8. 8
13	10. 6	262. 6	73	22. 8	295. 9	8	46. 4	281. 2	283	49. 7	9. 8
14	10. 0	263. 2	74	23. 3	296. 5	9	48. 2	282. 2	284	51. 5	10. 8
+15	−9. 4	263. 7	+75	+23. 9	297. 0	+10	+50. 0	283. 2	285	+53. 3	+11. 8
16	8. 9	264. 3	76	24. 4	297. 6	11	51. 8	284. 2	286	55. 1	12. 8
17	8. 3	264. 8	77	25. 0	298. 2	12	53. 6	285. 2	287	56. 9	13. 8
18	7. 8	265. 4	78	25. 6	298. 7	13	55. 4	286. 2	288	58. 7	14. 8
19	7. 2	265. 9	79	26. 1	299. 3	14	57. 2	287. 2	289	60. 5	15. 8
+20	−6. 7	266. 5	+80	+26. 7	299. 8	+15	+59. 0	288. 2	290	+62. 3	+16. 8
21	6. 1	267. 0	81	27. 2	300. 4	16	60. 8	289. 2	291	64. 1	17. 8
22	5. 6	267. 6	82	27. 8	300. 9	17	62. 6	290. 2	292	65. 9	18. 8
23	5. 0	268. 2	83	28. 3	301. 5	18	64. 4	291. 2	293	67. 7	19. 8
24	4. 4	268. 7	84	28. 9	302. 0	19	66. 2	292. 2	294	69. 5	20. 8
+25	−3. 9	269. 3	+85	+29. 4	302. 6	+20	+68. 0	293. 2	295	+71. 3	+21. 8
26	3. 3	269. 8	86	30. 0	303. 2	21	69. 8	294. 2	296	73. 1	22. 8
27	2. 8	270. 4	87	30. 6	303. 7	22	71. 6	295. 2	297	74. 9	23. 8
28	2. 2	270. 9	88	31. 1	304. 3	23	73. 4	296. 2	298	76. 7	24. 8
29	1. 7	271. 5	89	31. 7	304. 8	24	75. 2	297. 2	299	78. 5	25. 8
+30	−1. 1	272. 0	+90	+32. 2	305. 4	+25	+77. 0	298. 2	300	+80. 3	+26. 8
31	0. 6	272. 6	91	32. 8	305. 9	26	78. 8	299. 2	301	82. 1	27. 8
32	0. 0	273. 2	92	33. 3	306. 5	27	80. 6	300. 2	302	83. 9	28. 8
33	+0. 6	273. 7	93	33. 9	307. 0	28	82. 4	301. 2	303	85. 7	29. 8
34	1. 1	274. 3	94	34. 4	307. 6	29	84. 2	302. 2	304	87. 5	30. 8
+35	+1. 7	274. 8	+95	+35. 0	308. 2	+30	+86. 0	303. 2	305	+89. 3	+31. 8
36	2. 2	275. 4	96	35. 6	308. 7	31	87. 8	304. 2	306	91. 1	32. 8
37	2. 8	275. 9	97	36. 1	309. 3	32	89. 6	305. 2	307	92. 9	33. 8
38	3. 3	276. 5	98	36. 7	309. 8	33	91. 4	306. 2	308	94. 7	34. 8
39	3. 9	277. 0	99	37. 2	310. 4	34	93. 2	307. 2	309	96. 5	35. 8
+40	+4. 4	277. 6	+100	+37. 8	310. 9	+35	+95. 0	308. 2	310	+98. 3	+36. 8

4

Construction Practices

Craftsmanship is more important in building oceanographic equipment than in building equipment for practically any other field of science. A poorly built instrument may work in a laboratory or school room, but it will fail or operate erratically in the rough ocean environment.

Most failures in oceanographic instruments can be traced directly to poor mechanical construction rather than to basic design. Particular care must be taken to ensure that parts are solidly mounted and components such as batteries are securely clamped down. It is inevitable that oceanographic instruments will be dropped and banged around. By building them to withstand these jolts, you will avoid the frustration and disappointment of a wasted trip caused by a damaged instrument.

The instruments and equipment described here can be built by using simple hand tools. A good basic tool kit would include:

- Screwdriver, flat blade, ¼-inch blade, 3-inch shank
- Screwdriver, Phillips medium 3-inch shank
- Diagonal cutters, 6-inch
- Soldering iron, 30-watt
- Rosin core solder
- Hand drill or ¼-inch electric drill
- Drill bits (metal) 1⁄32-, ⅛-, 1⁄16-, and ¼-inch
- Sturdy pocket knife
- Medium-coarse file
- Coping saw
- Hand reamer, ½-inch
- Alligator-clip or small clip-on heat sink
- Ruler with metric scale, 12-inch
- Small crescent wrench
- Adjustable wire strippers

The electronic components you need can be bought at most popular electronic stores or ordered by mail from mail-order part houses. If this is your first try at building an electronic instrument, it is recommended that you start by building a simple project, such as the electronic thermometer or anemometer.

Soldering correctly is the key to reliable electronics. Use rosin-core solder only! The acid-core solder found in plumbing and hardware stores will wreak havoc with any electronics project. Before starting, clean the tip of your soldering iron and tin it with solder. Make sure the tip is free from pits, using the file to dress it down and remove the pits if necessary. If you have not soldered before, practice for a while on some scrap wire until you become proficient. It will take only a few minutes.

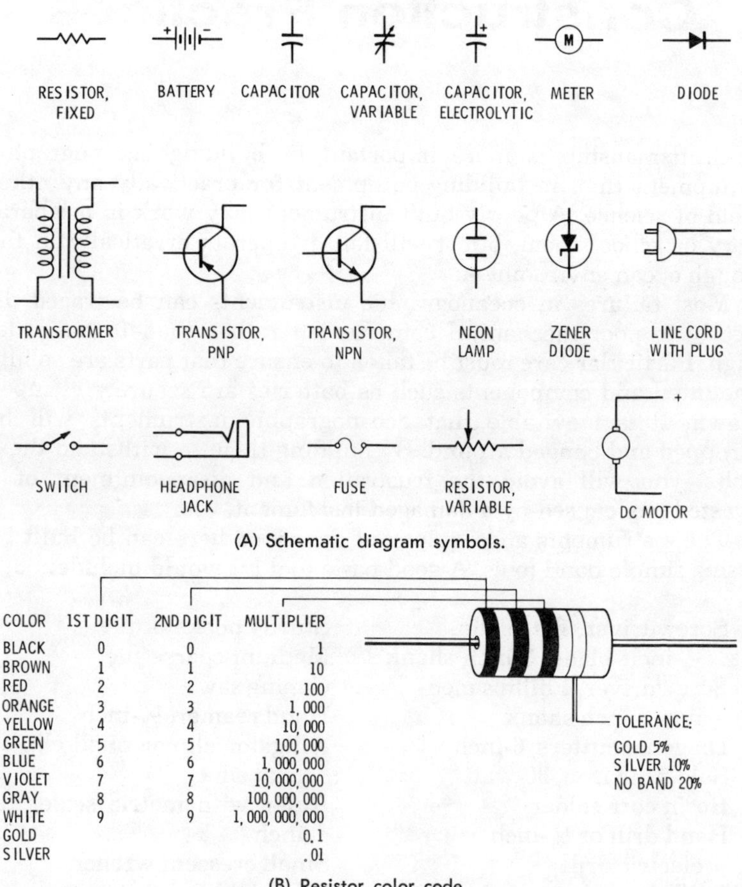

(A) Schematic diagram symbols.

COLOR	1ST DIGIT	2ND DIGIT	MULTIPLIER
BLACK	0	0	1
BROWN	1	1	10
RED	2	2	100
ORANGE	3	3	1, 000
YELLOW	4	4	10, 000
GREEN	5	5	100, 000
BLUE	6	6	1, 000, 000
VIOLET	7	7	10, 000, 000
GRAY	8	8	100, 000, 000
WHITE	9	9	1, 000, 000, 000
GOLD			0.1
SILVER			.01

(B) Resistor color code.

Fig. 4-1. Parts

In soldering components, a good mechanical connection is just as important as a good solder connection. Solder has very poor mechanical strength.

The leads to be soldered should be clean and free from corrosion. To solder, place the tip of the iron on the leads to be soldered, and hold the solder against the wires. When the wires become heated sufficiently, the solder will flow over them.

When soldering semiconductor devices such as diodes and transistors, clip a small alligator clip or heat sink between the point you are soldering and the body of the device. This will prevent damage to the component from the heat of the soldering iron.

Use only the minimum amount of solder necessary to flow over the connection. After completing the solder joint, inspect it carefully. A dull frosted appearance indicates a cold solder joint that

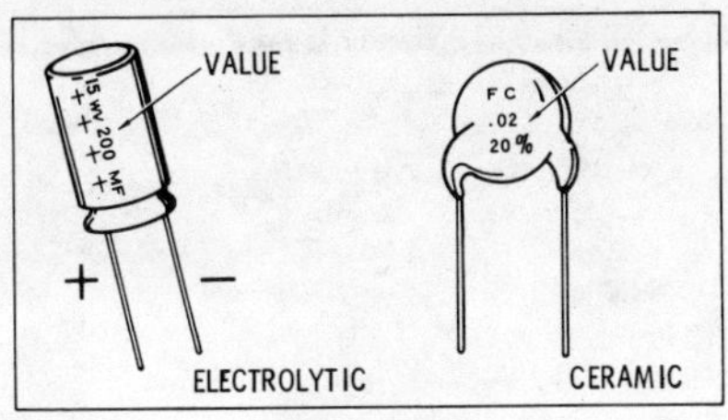

(C) Capacitors.

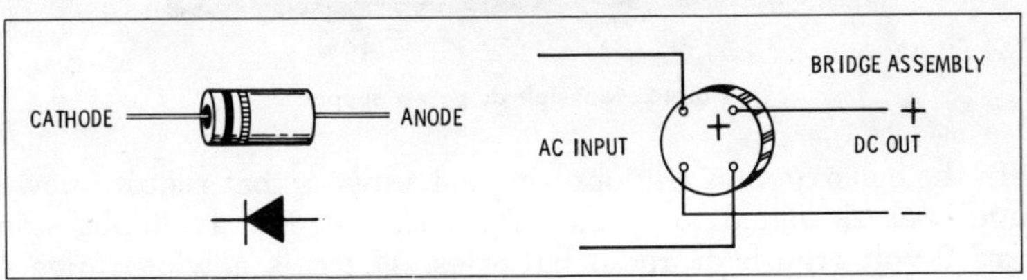

(D) Diodes or rectifiers.

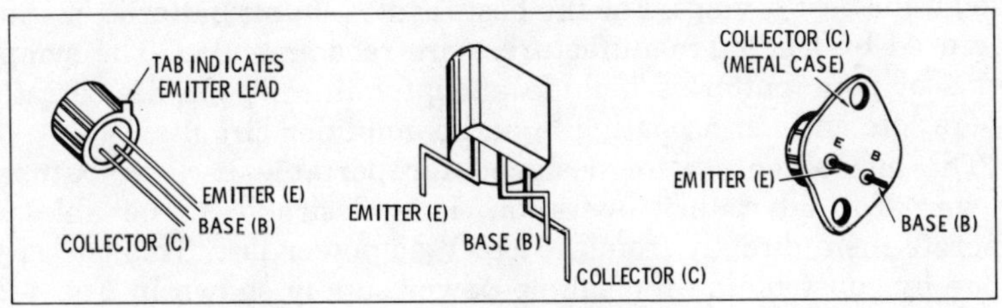

(E) Transistors.

identification.

will result in a poor electrical connection. Good solder connections are bright and smooth.

Components include resistors, capacitors, diodes, transistors, thermistors, meters, potentiometers, jacks, and switches. Their identification is shown in Fig. 4-1. Resistors are readily recognized by their shape. Their value is designated by colored bands. The first three bands from the left are used in reading the value of the resistor in ohms. The last band, either gold or silver, gives the tolerance of the resistor. Resistors having gold bands will have values within 5% of the value designated by the colored bands, while those with silver bands will have a 10% variation. Sometimes resistors are found with no tolerance band. These have a tolerance of 20%.

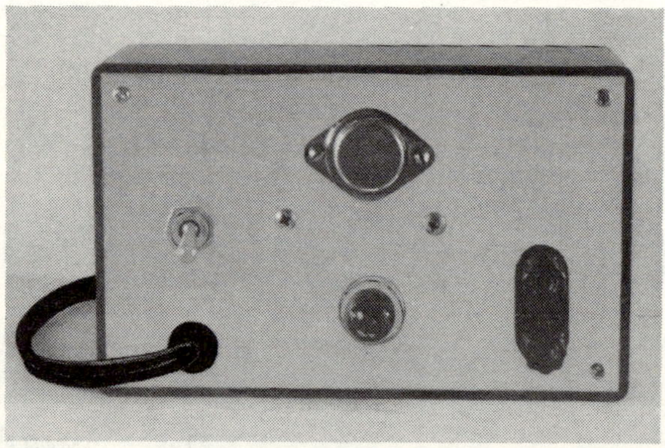

Fig. 4-2. Nine-volt dc power supply.

In the construction sections, the instruments that require power have been designed to operate from the readily available, standard 9-volt transistor radio batteries. There is a wide range of quality available in these batteries. The price of the cheaper ones makes them attractive, but they have disadvantages that outweigh the cost savings. For the best results, 9-volt batteries manufactured by leading manufacturers are recommended. The symptoms of poor batteries include abrupt failure, poor life, erratic operation, and motorboating in audio amplifier circuits.

The instruments were designed for portable use, on battery power. For laboratory operation, it is desirable to be able to operate them directly from the 115-V ac power line. This is easily done by constructing the simple power supply shown in Fig. 4-2. The schematic is shown in Fig. 4-3.

The transformer (T) steps the line voltage down from 115 volts to 12 volts. This alternating-current voltage is rectified by the bridge rectifier (CR) into direct current. This is filtered by the capacitor (C1), prior to being regulated to 9 volts by the power transistor (Q). The reference voltage for the regulating transistor is provided by the zener diode. To protect the regulating transistor from thermal damage, the transistor is mounted on the outside of the chassis box panel. Overload protection is provided by a ½-ampere slow-blow type fuse.

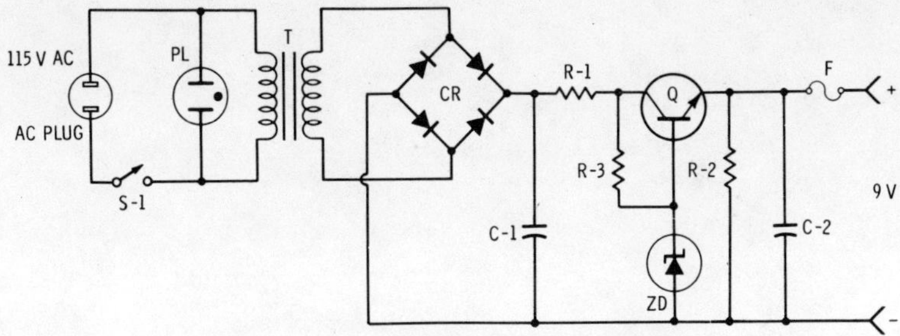

Fig. 4-3. Power supply schematic diagram.

Table 4-1. Power-Supply Parts List

Symbol	Quantity	Description
PL	1	Neon pilot lamp with built-in resistor, Dialco 95-0408 (or equiv)
T	1	Transformer, 115-V primary; 12-V, 200-mA secondary
CR	1	Bridge rectifier assembly, 5-A, 50 piv
C1	1	Capacitor, 2000-μF, 25-V
C2	1	Capacitor, ceramic .01-μF, 50-V
R1	1	Resistor, 2.4-ohm, ¼-watt
R2	1	Resistor, 330-ohm, ¼-watt
R3	1	Resistor, 1000-ohm, 1-watt
Q	1	Transistor, npn 2N389
ZD	1	Zener diode, 10.1-V, ¼-watt
F	1	Fuse, ½-Amp slow-blow
S1	1	Toggle switch, spst
Misc		
	2	Terminal posts
	1	Chassis box
	1	Line cord, ac

5

Instruments and Apparatus

Construction details for various oceanographic instruments that you can build are given in this chapter. Included are an electronic thermometer, current-speed and direction-measuring system, wind-speed and direction indicators, Secchi disc, underwater sound system, aquatic biotelemetry system, and a plankton sampling net.

ELECTRONIC THERMOMETER

Temperature measurements are among the oldest and most important measurements made by marine scientists. Seawater temperature is used to study the air-sea interface, help forecast weather, study deep ocean currents, and make fishery studies.

For many years the mercury thermometer was the only way to measure water temperature. Surface temperature was measured by taking a sample in a bucket and immersing a thermometer. Temperature at depth was measured by the reversing thermometers described in Chapter 1. The mercury and reversing thermometers are quite accurate, but expensive and easily broken. The electronic thermometer described here is very rugged, using a remarkable electronic device called the thermistor.

The thermistor has a very interesting characteristic. Its resistance is a function of temperature. If a small voltage is applied to a thermistor, as shown in Fig. 5-1, a current will flow. This current can be measured. It is proportional to the temperature. Of

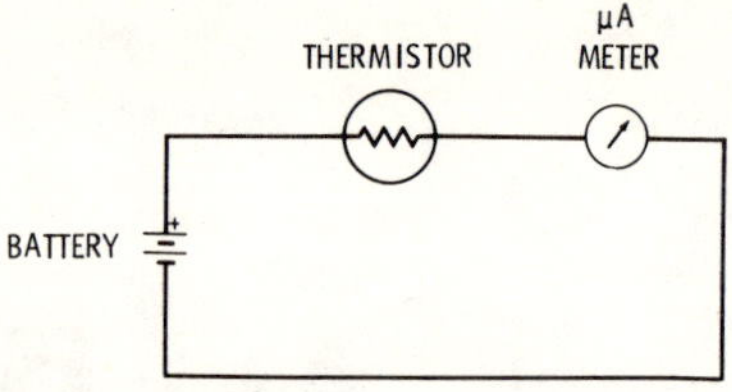

Fig. 5-1. Basic thermistor circuit.

course, the current must be kept quite small to prevent the thermistor from self-heating. To construct a thermometer, we simply build a circuit that will accurately measure the resistance of the thermistor, and then we calibrate the dial in terms of temperature. Fig. 5-2 shows the variation of resistance with temperature for a typical thermistor.

The circuit of the electronic thermometer is shown in Fig. 5-3. It is known as a Wheatstone bridge after its inventor, Sir Charles

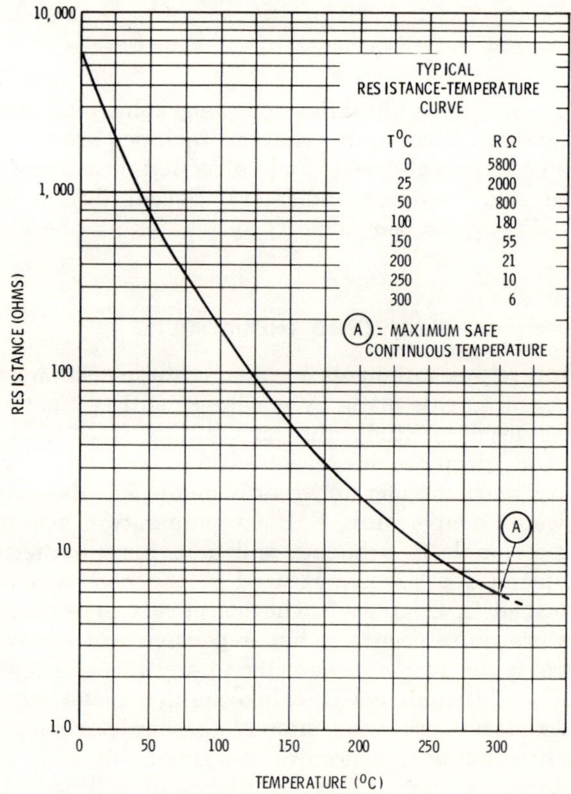

Fig. 5-2. Typical thermistor curve.

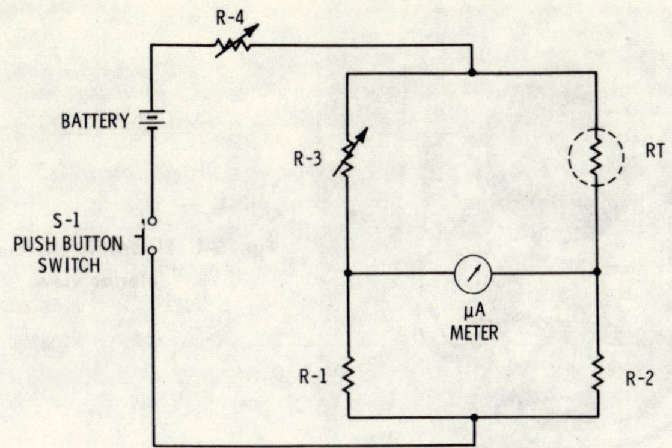

Fig. 5-3. Electronic thermometer circuit diagram.

Wheatstone. When the resistance in R1 is exactly equal to that of the thermistor, R-T, no current will flow through the meter.

The circuit uses a zero center meter. To make a measurement, R3 is simply adjusted until no current flows through the meter. The temperature is read from the dial. A 100-0-100 microampere meter is used as a compromise between ruggedness and sensitivity.

A meter requiring less current for full-scale deflection, say a 50-0-50 microampere meter, would provide greater sensitivity but could more easily burn out if the meter were turned on with the probe unplugged.

The meter portion is built in a standard Bakelite meter box. These boxes come complete with a Bakelite or aluminum panel. Do not remove the protective paper from the panel until you have drilled and cut all necessary holes. The hole for mounting the meter is easily cut by using a coping saw. All the components, including a clamp for holding the battery, are mounted directly to the panel. (See Fig. 5-4.)

Next the probe is built. Care must be taken not to damage the thermistor while soldering its leads to the cable. Fig. 5-5 shows how the probe is assembled. A small shield is made by cutting off a short piece (about 3 inches) of pvc (polyvinyl chloride) water pipe and drilling a few holes in it to permit the water to circulate. The purpose of this shield is to prevent damage to the fragile thermistor. A plastic vial such as the kind used for pills may be used instead of the pvc, but it is not as rugged. The shield is left off during calibration.

After assembling the probe, test it by plugging it into the meter and checking to make sure that you can zero the meter by adjusting R1.

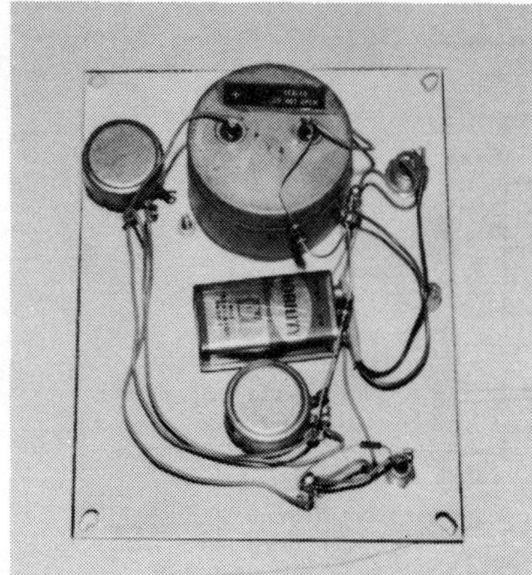

Fig. 5-4. Electronic thermometer—interior view.

Next, carefully waterproof the connections, using several coats of clear silicone rubber. Using the clear type permits you to inspect it later on for mechanical damage.

After the waterproofing rubber sealant has cured (usually 24 hours) the meter can be calibrated. First, cut a dial plate from a manila folder and cement it to the panel with rubber cement (Fig. 5-6). Calibration will require two beakers or water glasses, a mercury thermometer, water, salt, ice cubes, and a stirring rod. Start with the glass full of hot tap water. Dilute it with cool water until the temperature stabilizes at 40°C. Insert the thermistor probe into the water. Wait until the meter stops moving; then adjust it for zero, using R3. Mark this 40° point on the scale with a pencil.

Repeat this procedure every degree until zero degrees is reached. The water temperature is lowered by diluting the bath with cold water or crushed ice. With a little practice you can become quite proficient. To reach 0°C (the freezing point of water), common table salt must be added.

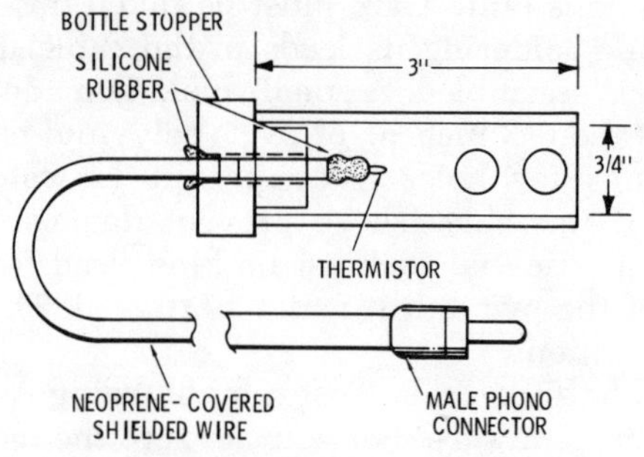

Fig. 5-5. Thermistor probe construction.

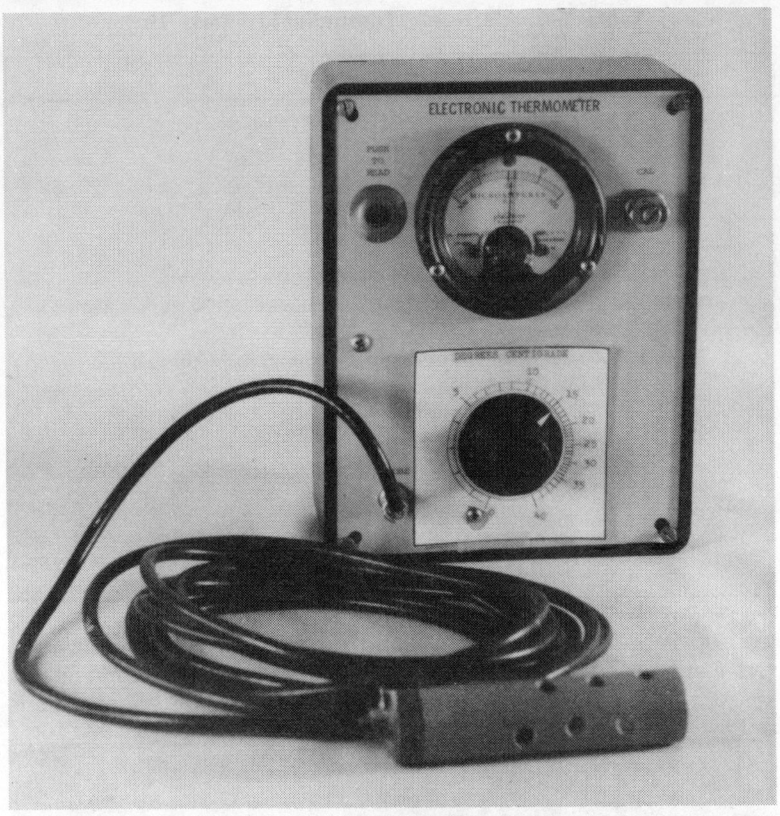

Fig. 5-6. Completed electronic thermometer.

After completing the initial calibration, recheck the accuracy of the scale markings by spot checking the calibration at ten-degree increments.

You will notice that the distance between degree marks widens as the temperature decreases. This is due to the logarithmic response of the thermistor. For example, a temperature change from 0 to 1 degree is equivalent to a resistance change of 810 ohms, while a change in temperature from 39 to 40 degrees is equivalent to a resistance change of only 110 ohms.

Table 5-1 is the parts list for the electronic thermometer.

CURRENT-SPEED AND DIRECTION-MEASURING SYSTEM

The current-speed and direction-measuring system (Fig. 5-7) is built around a common electronic indicator unit that connects to

Table 5-1. Electronic Thermometer Parts List

Quantity	Description
1	Bakelite utility case with aluminum panel, 7 3/4″ × 4 7/16″ × 2 3/8″
1	Meter, 100-0-100 microampere
2	Resistors R1 and R2, 500-ohm, ¼-watt
1	Battery, 9-volt (transistor radio type)
1	Battery connector
1	Switch, push-button (Switchcraft 3501FP or equiv)
1	Chassis-type female phonograph connector (Switchcraft 3501FP or equiv)
1	Male phonograph connector (Switchcraft GC770 or equiv)
–	Cable (Belden 8401 or equiv)
–	Silicone rubber sealant (General Electric RTV-108 or equiv)
1	Thermistor (Fenwal JB31J1 or equiv)
–	Bottle stopper
–	¾″ pvc pipe, 3″ long, or plastic pill bottle
1	Knob (to fit ¼″ shaft)
1	Resistor, R4, 4000-ohm
1	Resistor, R3, 10,000-ohm, variable

the separate sensors for speed and direction. Neoprene-jacketed two-conductor (AWG No. 18) power cable is used to connect the sensors to the terminals on the indicator unit. The equipment is

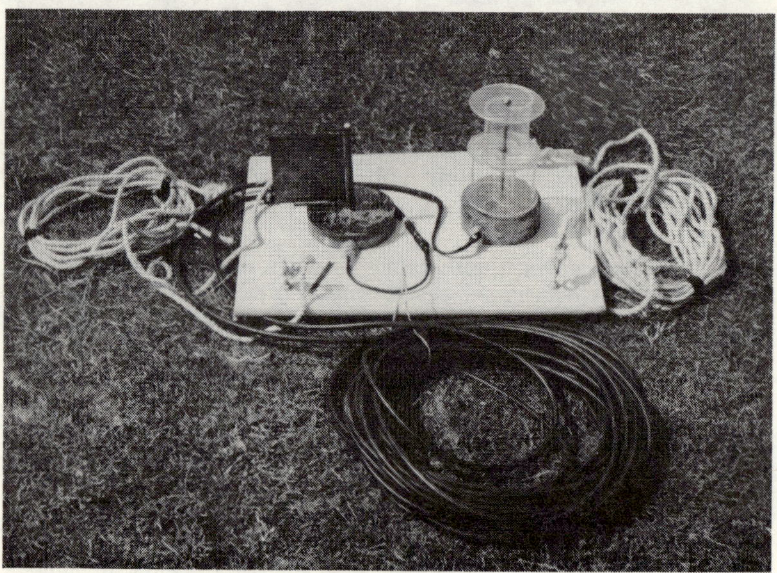

Fig. 5-7. Current speed and direction sensors.

capable of measuring currents in the range of 0.5 to 5 knots, and direction in 45-degree increments.

The indicator unit, shown in Fig. 5-8, is constructed in a homemade wooden case measuring 8 by 13 by 3 inches. All the components are mounted to the panel which is fastened to the case with wood screws. The circuit (Fig. 5-9) consists of three separate measuring circuits using a common meter for the indicator. In the battery test position, the meter tests the condition of the internal battery. In the direction position, the instrument functions as an ohmmeter to measure the resistance of the elements of the direction sensor. In the current-speed position, a monostable, or "one shot," multivibrator acts as a tachometer and counts pulses generated by the closure of a magnetic reed switch in the base of the current-speed sensor.

Fig. 5-8. Portable indicator unit.

The mechanical construction and assembly details of the sensors are shown in Figs. 5-10 and 5-11. The most important parts of both sensors are small permanent magnets and their companion magnetic switch elements. These components are widely available at electronic hobby stores. The circuit of the direction sensor is shown in Fig. 5-12. Common ¼-watt carbon resistors are arranged in 45-degree increments and placed in series with the reed switch elements. As the water current causes the vane to move, the small magnet cemented to the edge of the vane moves over the switch elements, closing them one at a time. This connects the appropriate resistor into the ohmmeter circuit. The meter scale is calibrated in compass points instead of ohms. Table 5-2 is the parts list for the speed and direction sensors.

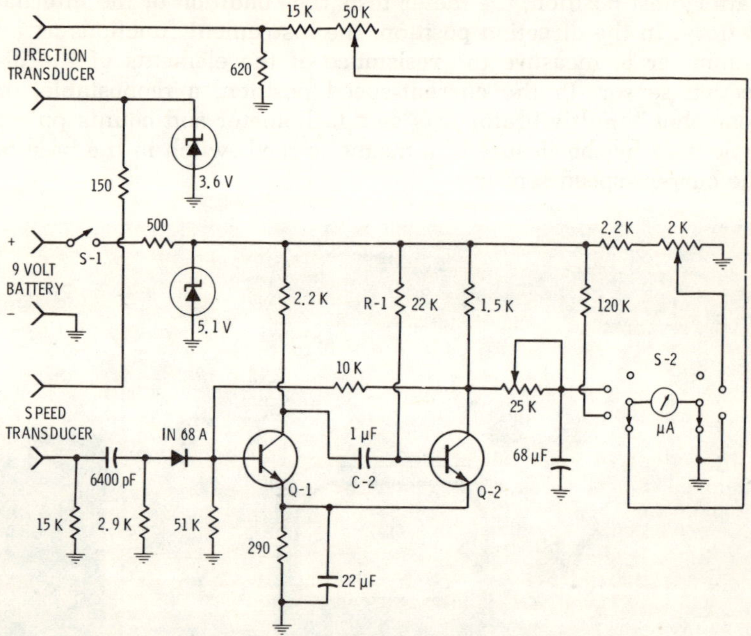

Fig. 5-9. Current speed and direction indicator schematic diagram.

The current-speed sensor (Fig. 5-13) is built from the same type of plastic sheet as the direction sensor. The small magnets are cemented to the bottom side of the spacer closest to the base. Again, use of a protractor is essential. Failure to accurately space the magnets will result in erratic operation. The magnets should be carefully spaced at 45-degree intervals and located on the radials so that they will pass directly over the single-switch element in the base as the rotor turns. Magnets must be mounted with alternate poles facing the rim. They can be glued on satisfactorily with epoxy glue. However, a more professional job results if they are inletted into the plastic as was done with the resistors and reed switches for the direction sensor.

Plastic parts are easily cut using a coping saw or jigsaw. The rotor vanes are formed by heating the plastic pieces in an oven at

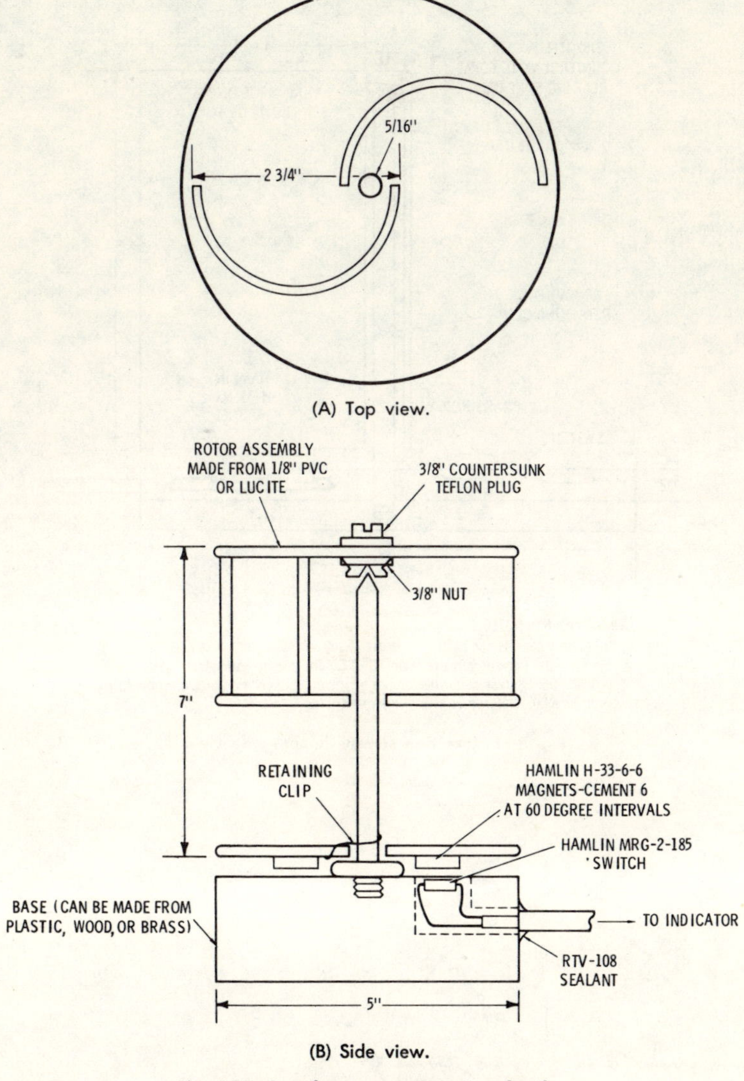

Fig. 5-10. Speed sensor construction details.

250°F until they are soft and pliable, then forming them around a tin can. (Note: The room *must be well ventilated.*) If pvc thin-walled pipe is available, the rotor vanes can be made from short sections cut in half.

The assembled rotor is slipped over the shaft and retained in place by the spring wire retainer. The tension on the retaining

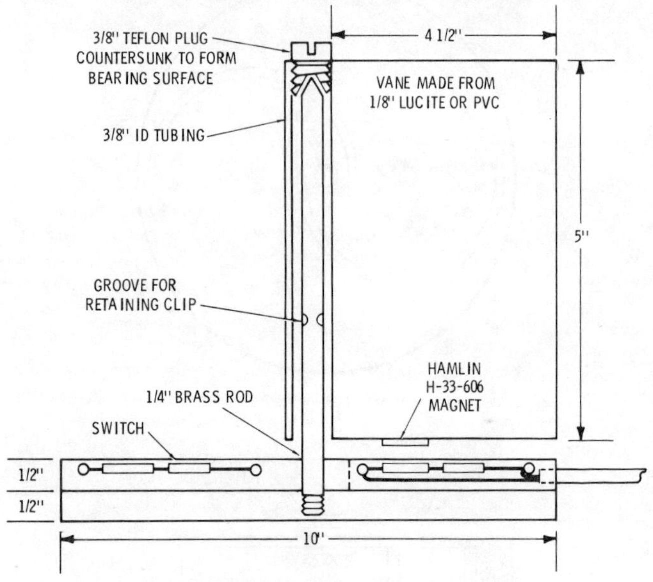

CONSTRUCTION NOTES:

1. Assembly can also be built from brass sheet and tube instead of plastic.
2. Be sure the cement you use is designed for the plastic you select.
3. A BNC (UG-88) type of connector can be installed to permit quick disconnecting of the cable. Waterproof with shrink on tubing.

Fig. 5-11. Direction sensor details—side view.

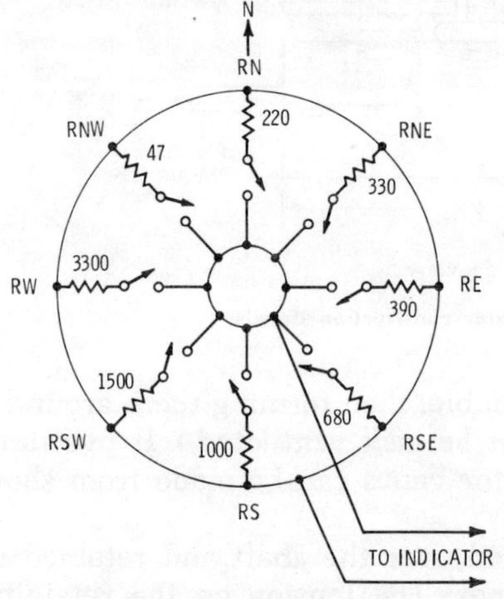

Fig. 5-12. Direction sensor circuit.

Table 5-2. Speed and Direction Sensors Parts List

Quantity	Location	Description
		Direction Sensor
1	RN	Resistor, 220-ohm
1	RNE	Resistor, 330-ohm
1	RE	Resistor, 390-ohm
1	RSE	Resistor, 680-ohm
1	RS	Resistor, 1000-ohm
1	RSW	Resistor, 1500-ohm
1	RW	Resistor, 3300-ohm
1	RNW	Resistor, 47-ohm
8	—	Reed switches (Hamlin MRG-2-185 or equiv)
1	—	Magnet (Hamlin H-34-607 or equiv)
		Speed Sensor
1	—	Reed Switch (Hamlin MRG-2-185 or equiv)
6	—	Magnets (Hamlin H-34-607 or equiv)

Note: There are a wide variety of reed switch elements and magnets available in electronic component and hobby stores. Practically any combination can be accommodated simply by making the mounting slots the appropriate size. Switch elements should be single-pole-single-throw **form A** configuration.

spring should be adjusted so that it is as light as possible in order to minimize drag.

The assembly is tested by connecting to an ohmmeter and slowly watching the meter to verify that the switch closes as each magnet passes over the switch element. Failure of a switch element is rare. Faults are most likely to be found in magnet placement or failure to alternate magnet poles. After testing, the cable hole is filled with silicone rubber.

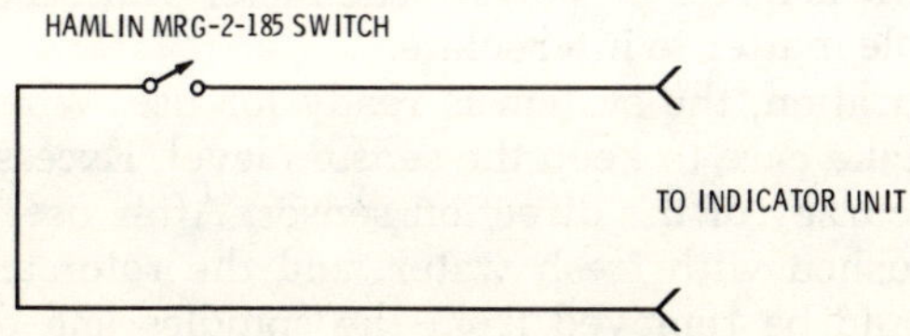

Fig. 5-13. Speed sensor circuit.

The direction sensor (Fig. 5-11) is constructed on a base made from ½-inch-thick pvc or Plexiglas. Nonferrous metals such as brass can also be used. The radials should be laid out carefully with a protractor. The north reference should be clearly and permanently marked.

A ¼-inch drill or small burr is used to cut the grooves for inletting the resistors and switch elements into the plastic. This is best done with a powered hand grinder. After inletting, the sensor is wired. Then the neoprene two-conductor cable is soldered in place. The sensor is tested by connecting the leads to an ohmmeter or the indicator unit, and then manually moving the vane and observing the scale. The correct resistance value should be shown as the magnet passes over the respective reed switches.

The grooves and the cable exit hole are then packed with RTV-108 silicone rubber sealant. The remainder of the sensor is assembled after the rubber has cured. The supporting rod is mounted to the base, and the vane assembly slipped over it. The small retaining clip, made from a short length of stainless steel spring wire, is inserted.

The system is now ready for checkout and calibration. It is convenient to mount both sensors on a small platform with eyebolts for handling lines and a prominent mark for orienting to north (see Fig. 5-7). Connect the sensors to the indicator. Turn the function switch to the battery test position. With a fresh battery installed, the meter should read between 80% and full scale. Next, switch the function switch to the speed position. Adjust R2 until the meter reads zero. Turn the Savonius rotor rapidly by hand and observe that the meter needle moves up scale. Switch to the direction position and move the vane clockwise from the north reference mark. Mark the compass points on the meter scale as the successive switches close.

The current-speed meter has to be calibrated before it can be used. One way is to find a stream and submerge the sensor. Using a stopwatch to time small chips of wood floating past a known distance, time the current. Calculate the speed: distance divided by time. Repeat the procedure at different points along the stream where the flow is faster or slower. The meter scale is quite linear, so it is a simple matter to interpolate.

After calibration, the system is ready for use. When operating the system, take care to keep the sensors level. Excessive tilt will affect the accuracy of the direction sensor. After use, the sensors should be flushed with fresh water, and the rotor and vane assemblies should be removed from the spindles and dried before storing.

WIND-SPEED AND DIRECTION INDICATORS

The meteorological measurements of most interest to oceanographers are air temperature, wind speed, and wind direction. Measuring air temperature is quite simple, since the electronic thermometer is perfectly suited for measuring air temperature in the range of 0 to 40°C as well as water temperature.

Fig. 5-14. Anemometer.

Wind speed and direction can be measured with rather simple equipment. Although sophisticated solid-state devices are now available for making these measurements, the familiar cup and vane are still the most commonly encountered instruments. Since measurements must be made at various locations, the equipment must be portable.

Anemometer

Fig. 5-14 shows the hand-held anemometer. It is built by using a small permanent-magnet toy motor available in most hobby shops. Three small cups are arranged at 120° increments around the shaft of the motor (Fig. 5-15A). The output of the motor (Fig. 5-15B) is surprisingly linear when used as a generator. As the wind turns the shaft, the motor acts as a dc generator. The meter is simply a dc voltmeter. The schematic diagram is shown in Fig. 5-16.

The cup assembly is made from plastic fishing floats (ping pong balls can be used) cemented to pieces of ⅛-inch-diameter plastic rod. Scraps from model kits can be used. The hub is a small Erector-set wheel. A nut is soldered to the setscrew to make it easier to install and remove the cup assembly. After assembling the anemometer, hold it out in the wind to test it. If the needle does not move up scale, simple flip the cup assembly over.

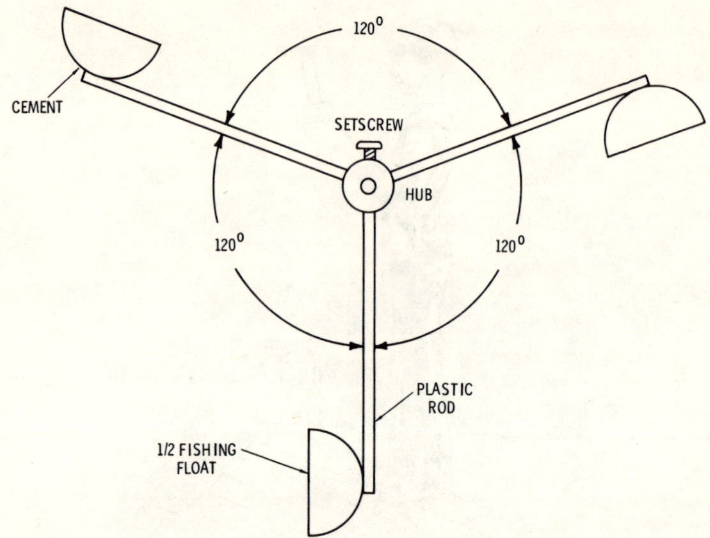

(A) Cup arrangement—top view.

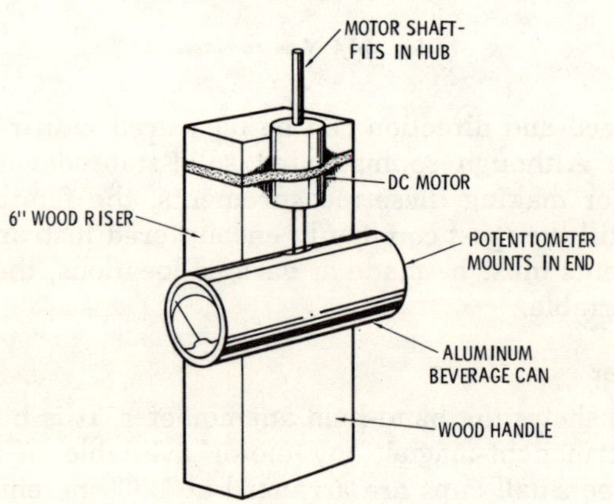

(B) Component placement.

Fig. 5-15. Anemometer construction details.

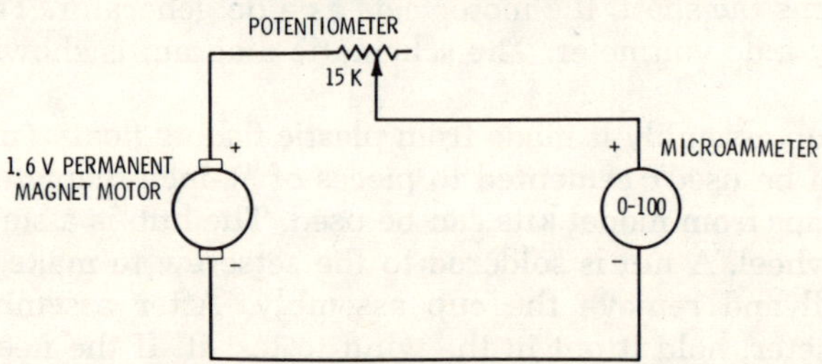

Fig. 5-16. Anemometer schematic diagram.

The anemometer is calibrated by comparing it with an automobile speedometer. One word of caution, do not try to drive and calibrate at the same time! Have someone else do the driving. Start off by driving at 50 mph. Adjust the potentiometer for full scale. Then, reduce the vehicle speed in ten-mile increments, logging the meter reading for each speed. Use this data to make a calibration chart which can be cemented on the handle.

Wind Vane

The wind vane is a simple instrument constructed from ¼-inch dowel, a tenpenny nail, a broomstick, a square of plywood, some ¼ hex nuts, and a short piece of brass rod. Construction details are given in Fig. 5-17. The vane section is made from aluminum flashing or stiff cardboard.

To construct the vane, notch the dowel (A) to accept the metal fin (B). Drill two holes as shown to secure the fin with two 4-40 machine screws, lock washers, and nuts. The bearing (C) is made from a one-inch piece of ⅜-inch brass or iron rod. Drill it as shown to accept the dowel. A ½-inch-deep hole is drilled at right angles to the dowel hole. The tenpenny nail fits into this hole to form the bearing. The dowel, with the fin attached, is slid through

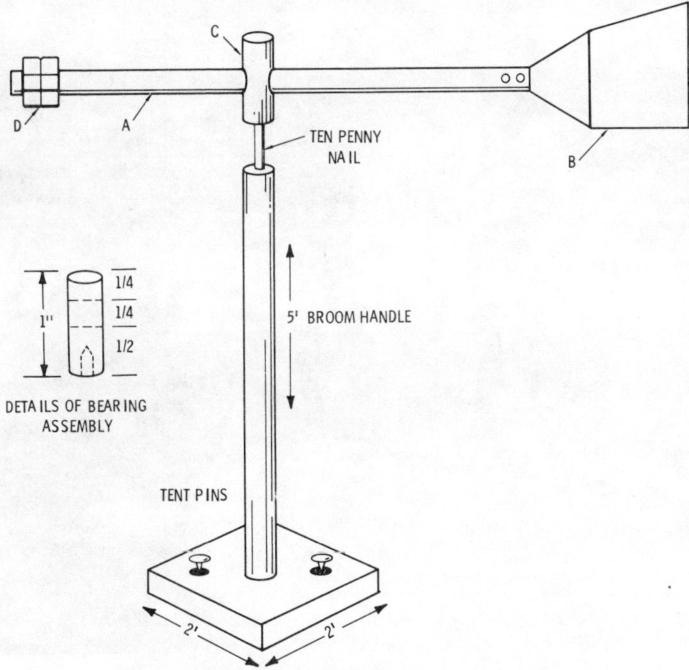

Fig. 5-17. Wind vane construction details.

(C) to its midpoint, then cemented in place. The hex nuts are forced onto the tip of the dowel and adjusted to balance the assembly.

The wind vane is now ready for use. A compass, preferably a lensatic type, is used to site along the vane, for determining direction accurately. Wind direction is always recorded as the direction from which the wind is coming.

TRANSPARENCY AND THE SECCHI DISC

The Secchi disc is a simple tool used to measure the transparency of seawater. It consists of a white disc about 30 centimeters (11.8 inches) in diameter.

Fig. 5-18. Using Secchi disc.

Although more-sophisticated electronic devices have been devised, the Secchi disc is still the most popular means of measuring transparency (Fig. 5-18). This is because of its low cost and simplicity—two important features for practical oceanographers. The materials used are listed in Table 5-3.

Table 5-3. Secchi Disc Parts List

Amount	Item
1	Plywood, 3⁄8″, 12″ × 12″
1	Eyebolt, 3⁄4″ eye, 1⁄4″ shank, 2″ long
2	Nuts, 1⁄4″ (for the above eyebolt)
1	Washer, flat 1⁄4″
1	Washer, lock 1⁄4″
75′	Line
	Wood sealer
	White enamel
	Scrap iron plate for weight

Cut the disc from the plywood, using a jigsaw or saber saw. Drill a 1⁄4-inch hole in the center. Next, apply at least three coats of good-quality marine wood sealer, allowing ample drying time between coats. After the sealer has thoroughly dried, apply two coats of white enamel. Assemble the hardware as shown in Fig. 5-19. Knot the line at every meter and paint the knots with international orange paint to make it easy to count them under water. To be really professional, splice the line around the eyebolt, using an eye splice rather than simply tying it.

The Secchi disc is now ready for testing (Fig. 5-20). The transparency or ability of water to transmit light depends upon a number of factors, such as the number, size, and nature of the particles

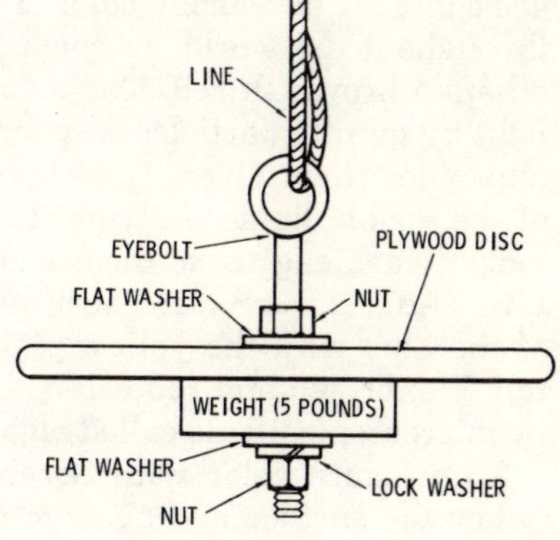

Fig. 5-19. Secchi disc construction details.

suspended in the water, as well as the nature and intensity of ambient light. Knowing how far that light penetrates is important to biologists because it reveals the depth of the euphotic zone, the area where photosynthesis is possible and the tiny phytoplankton change chemical nutrients into the primary food of the oceanic food chain. The amount of light is also important to divers and underwater photographers. They use Secchi discs to determine if there is enough light available for their tasks.

Fig. 5-20. Completed Secchi disc.

To use the Secchi disc, slowly lower it, over the side of a boat or off a pier until it just disappears from sight. Then count the number of knots by which it was lowered, and this will give you the depth. In some areas, such as badly polluted harbors or turbid coastal waters, it will disappear in a few meters. With adequate light it should be visible in clean coastal waters from 5 to 25 meters. The greatest recorded depth at which the disc has disappeared was 66 meters in the Sargasso sea.

In addition to measuring transparency, the Secchi disc is a good basic tool for measuring color. The color of seawater varies widely throughout the world. In some places it is a deep indigo blue, in others a brownish red. The color is due to the scattering of sunlight by minute particles suspended in the water and by the water molecules themselves. Blue light is at the short wavelength end of the visible light spectrum. It is more easily scattered than the longer wavelengths, so the ocean nominally appears blue. The yellowish-green color often seen near shore is caused by a mixture of the blue with the yellow pigment associated with phytoplankton. The brown and reddish-colored waters are frequently caused by microscopic plants called algae.

To measure color with Secchi disc, lower the disc one meter below the surface and observe the color of the water against the white disc. For best results you should have a standard for comparison, such as the Forel-Ule color comparator kit which can be obtained from scientific supply houses.

UNDERWATER SOUND SYSTEM

Due to the nature of seawater, virtually all studies rely upon instrumentation rather than direct observation. Light and radio (electromagnetic) waves are drastically attenuated by water while sound waves are not. Therefore, sound becomes a significant tool for probing the ocean depths. These depths are not silent. They abound in sounds ranging from those produced by marine animals to man-made noises from marine engines and sonar.

The basic instrument for studying these underwater sounds is the hydrophone or underwater microphone (Fig. 5-21). A hydrophone system (Fig. 5-22) consists of an acoustic transducer, an audio amplifier, and a headset or loudspeaker. Transducers employ a variety of principles. The most common are piezoelectric, electrostrictive, magnetostrictive, and variable reluctance. Commercial transducers are expensive. Occasionally, transducers from sonobuoys can be found on the surplus market for a few dollars.

An alternate approach is to make a variable-reluctance transducer. This is not a difficult project. Total cost for parts is only a

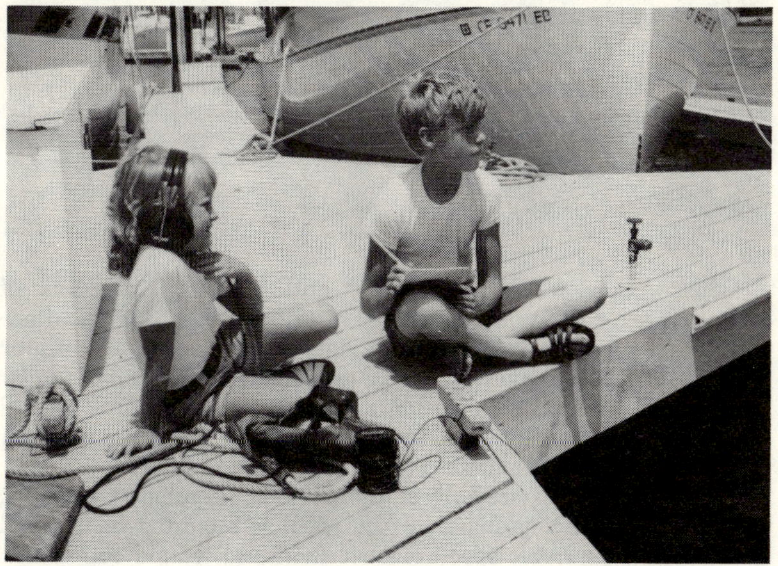

Fig. 5-21. Using simple hydrophone.

few dollars. The high-gain audio amplifier can be built for about six dollars.

A variable-reluctance transducer is similar in principle to the common permanent-magnet loudspeaker. Instead of the familiar paper cone, thin sheet-metal discs are used (Fig. 5-23). Sound waves strike the metal discs, resulting in a change in the magnetic field, which induces an alternating current into the coil. The resulting weak signal is amplified by the amplifier to a level suitable for headphones or a small 16-ohm loudspeaker.

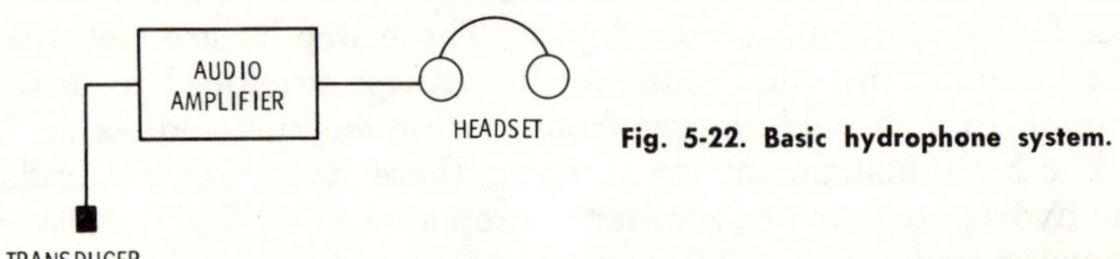

Fig. 5-22. Basic hydrophone system.

Sounds produced by marine organisms range in frequency from a few hertz to over 200 kHz. The heaviest concentration of sounds lies in the band of 100 Hz to approximately 800 Hz, well within the response of this transducer.

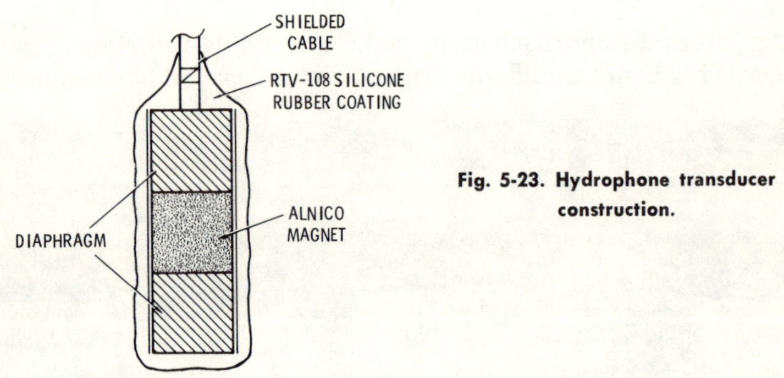

Fig. 5-23. Hydrophone transducer construction.

Construction of the transducer is quite simple. Dimensions of the parts are shown on Fig. 5-24. The end plates of the spool are cut with a jigsaw from ⅛-inch Plexiglas or pvc sheet. The center bushing is cut from pvc water pipe or is made from a pill bottle by cutting off the ends. The inside diameter of the bushing should be small enough to provide a snug fit for the alnico magnet. A good source of the magnet is a scrap loudspeaker.

After the spool is assembled, the parts are cemented together. Before the magnet is inserted, the spool is mounted on a small hand drill or variable-speed electric drill and wound full of No. 34 (American Wire Gauge) insulated magnet wire. Approximately 5000 turns provides adequate sensitivity. A fifty-foot length of RG-174/U miniature coaxial cable or neoprene-covered microphone cable is attached to the coil and terminated in a male phonograph connector. Strain relief at the coil is provided by wrapping a couple of turns of lacing twine around the coil, then around the shank of the transducer cable. Continuity of the coil should be checked with an ohmmeter.

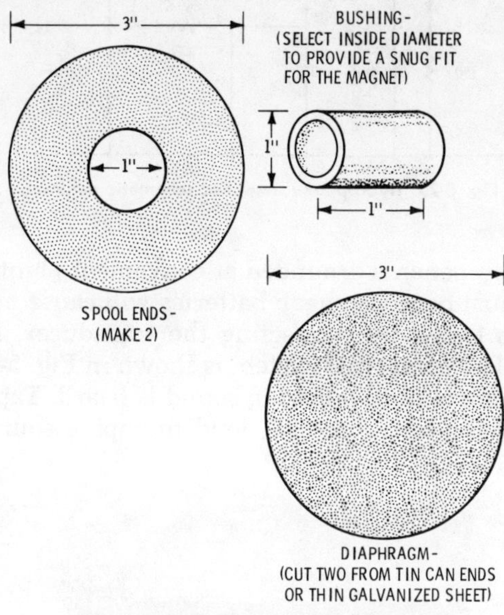

Fig. 5-24. Transducer parts dimensions.

Place the end plates on the spool. They will be held securely by the magnet. Liberally coat the entire assembly with General Electric RTV-108 silicone rubber. Apply this in several thin coats, using an artist's steel spatula. Be careful not to get the silicone rubber into your eyes as it can be dangerous. Allow the rubber to cure for at least 24 hours. Inspect for pin holes and cover up any holes with more silicone rubber.

Assemble the electronics on a 2½-by-5-inch piece of perforated board. Mount the completed electronics on the 3½-by-6-inch aluminum panel, using 1½-inch standoff bushings. The circuit diagram (Fig. 5-25) has no critical parts. Layout of the amplifier is shown in Fig. 5-26. Amplifier gain is quite high, so precautions should be taken to prevent unwanted oscillations.

Battery leads must be kept as short as possible and must be at least No. 20 (AWG) wire. A common ground wire must be used

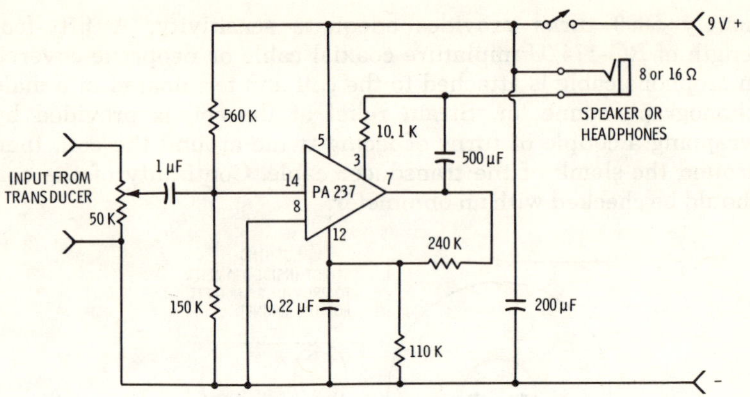

Fig. 5-25. Hydrophone amplifier schematic diagram.

rather than components grounded at convenient points. Good batteries are a must because cheap batteries will cause *motorboating.*

The unit is tested by connecting the transducer, battery, and headphones. The completed system is shown in Fig. 5-27. Advance the volume control until a hissing sound is heard. Tap your finger on the side of the transducer. A loud thumping sound should be clearly heard.

Fig. 5-26. Hydrophone amplifier parts layout.

Although the audio output is adequate to drive a 16-ohm speaker, headphones will produce the best field results. Standard 8- or 16-ohm headphones should be used.

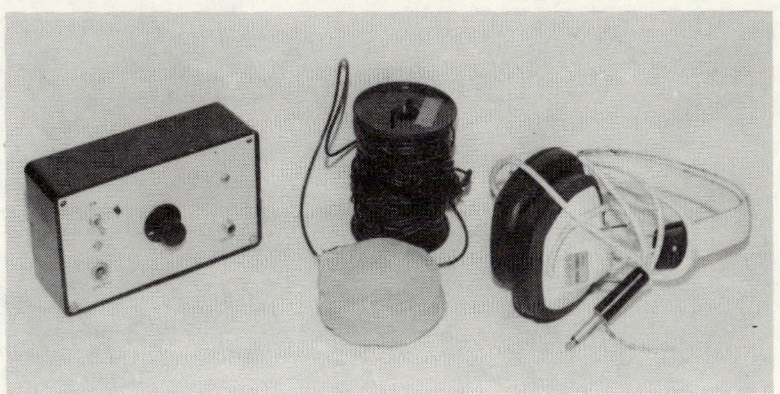

Fig. 5-27. Complete hydrophone system.

AN AQUATIC BIOTELEMETRY SYSTEM

The value of telemetry, the science of measuring at a distance, has been well recognized in missile and space work. The principles can also be used underwater. Aquatic biotelemetry, or marine biotelemetry as the science is more frequently called, works in a more hostile environment than space telemetry systems. The information must be transmitted over an acoustic link since radio frequencies are rapidly attenuated in water.

Even with this constraint, valuable information can be retrieved from marine animals, over surprisingly long distances. The information can range from simple directional tracking, or

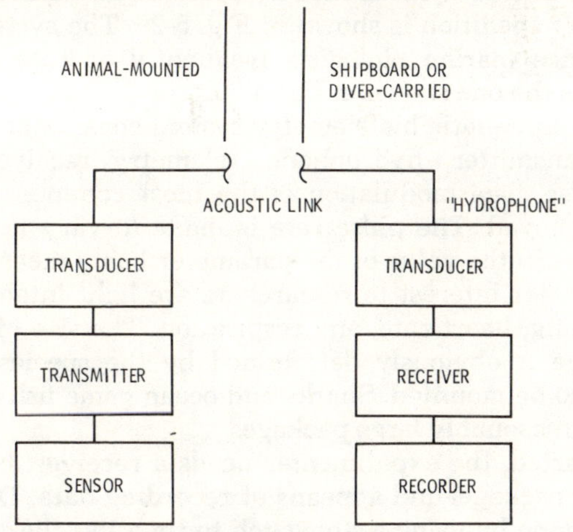

Fig. 5-28. Basic marine telemetry system.

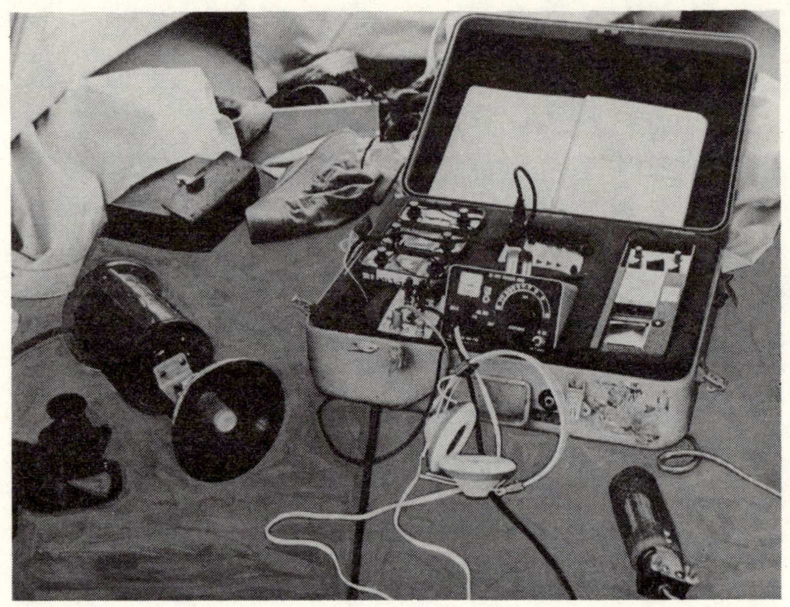

Courtesy E. A. Standora

Fig. 5-29. A portable biotelemetry system used by marine biologists in shark tracking studies.

"fish tagging," to multichannel monitoring of a marine animal's bodily functions.

A block diagram of a basic marine biotelemetry system is shown in Fig. 5-28. The equipment used by professional marine biologists while on an expedition is shown in Fig. 5-29. The system must be compact since marine biologists frequently operate from small craft such as the one shown in Fig. 5-30.

Basically an aquatic biotelemetry system consists of a sensor or sensors, transmitter, hydrophone, telemetry receiver, and recorder. Pulse code modulation is the most common modulation scheme employed. The pulse rate is made to vary in proportion with changes in the value of the parameter being measured. Parameters of major interest to researchers are light intensity, depth, speed, heading, heart rate, and respiration. The size of the telemetry package is obviously determined by the species of fish on which it is to be mounted. Sharks and ocean game fish are capable of carrying reasonably large packages.

To get started, the experimenter needs a receiver, hydrophone, transmitter package, and a means of recording data. Data recording can be done by using a stopwatch to time the clicks and a pad to write down the data.

Courtesy E. A. Standora

Fig. 5-30. Typical marine biologist's work boat.

An alternative method of data recording is to use a battery-operated cassette recorder to record the clicks in the field and to later reduce the data by hand in the laboratory. The recorder can be played back into a strip-chart recorder for graphic analysis.

Receiver

The receiver should be built first since it will be used as a piece of test equipment for checking out the transmitter package. The receiver (Fig. 5-31) is designed to be tunable from 30 to 70 kHz, a popular frequency range for biotelemetry. It is a type known as a *homodyne*. The mixing frequency differs from the incoming signal frequency by an amount that will produce an audible beat note or "click" in the headset. The schematic is shown in Fig. 5-32. An n-channel FET is used for the rf amplifier stage. The input transformer is an 88-mH telephone-type toroidal coil. This type of coil

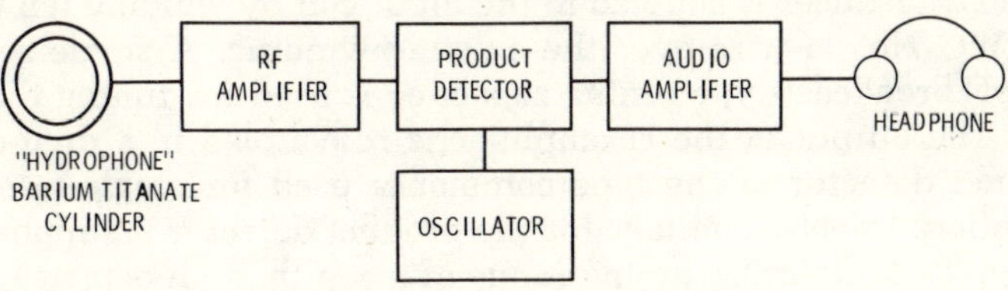

Fig. 5-31. Receiver block diagram.

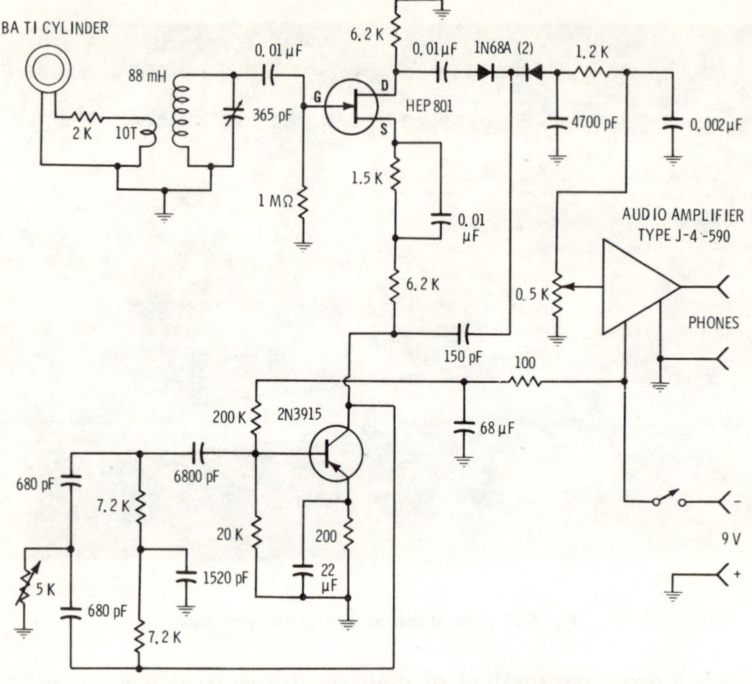

Fig. 5-32. Biotelemetry receiver schematic diagram.

is widely available in surplus electronic component houses. The transducer element (the same type of element is used for both the transmitter and the receiver) was obtained from Channel Industries in Santa Barbara, California. This company manfactures a complete line of transducer elements both in barium titanate and lead zirconate titanate materials. The ones used here are the hollow-cylinder barium titanate types, part number 1632. The cylinders are silver plated and come without the leads attached. Care must be taken when attaching the leads to prevent cracking the cylinders! Use as little heat as possible to minimize the chance of depolarizing the cylinder.

Transducer

The transducer is coupled to the input coil by winding ten turns of AWG No. 24 wire over the existing winding. A single section 365-pF broadcast-type tuning capacitor is used for tuning the circuit. The output of the rf amplifier is rc coupled to a dual-diode product detector of the type commonly used for single-sideband reception. Injection voltage for the product detector is supplied by a twin-T oscillator. A tuning range of more than an octave is possible with this circuit. Although a standard 5K potentiometer may be used for tuning, a ten-turn helical potentiometer of the same value will provide excellent bandspread. In constructing the twin-T circuit, remember that the transistor must have high gain to ensure reliable operation.

Audio Amplifier

The audio amplifier is a packaged one-watt unit having a 10K input impedance. The entire receiver is assembled on a piece of perforated fiberglass board. With the receiver made waterproof and the transducer mounted in a directional housing, the unit can be used underwater by scuba divers in recovering lost packages and for tracking animals. Mechanical details of the housing are provided in Figs. 5-33 and 5-34. The housing was built from pvc pipe with polystyrene end plates. The end-plate mounting for the

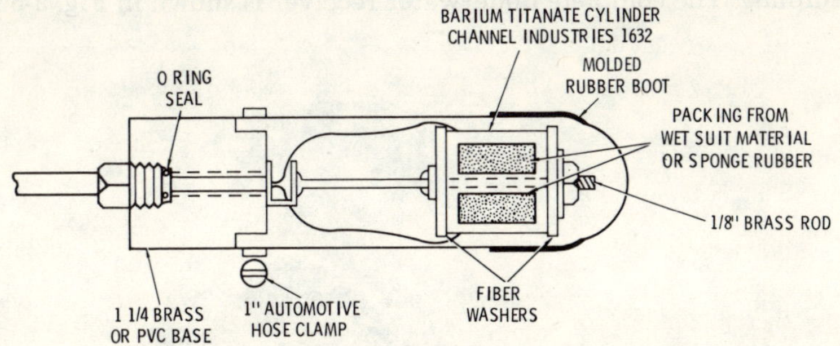

Fig. 5-33. Hydrophone assembly mechanical details.

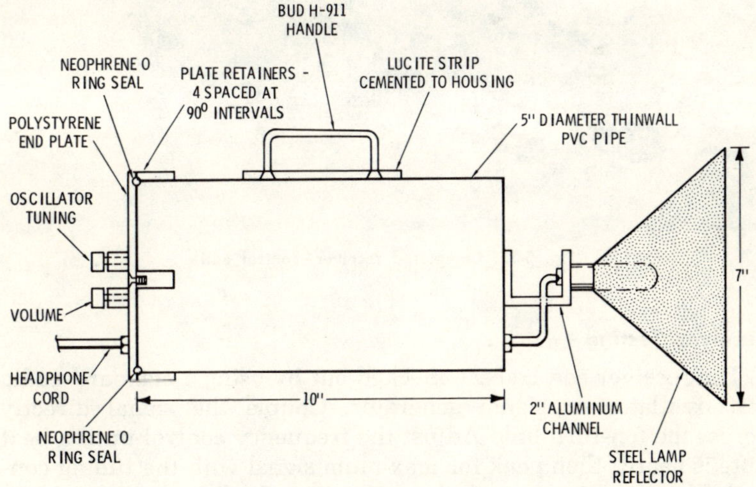

Fig. 5-34. Receiver housing mechanical details.

transducer is cemented in place, while the other one is made removable with an O-ring seal fitted to ensure against leakage.

Earphone

The earphone can be a commercial unit, such as those made by Pacific Divers, or can be made from a surplus ANB-H-1 headphone unit from a surplus HS-33 headset. To modify the headphone unit, carefully drill a 1/4-inch hole through the back of the case to permit pressure equalization. Next, coat the coils and cord connections with RTV-108 silicone rubber cement to waterproof them. A headband can be purchased from a swim and dive shop. Be sure to use two-conductor neoprene-covered cable. After each use, the cap should be unscrewed and the diaphragm removed. Flush the unit thoroughly with fresh water and dry before reassembling. The complete underwater receiver is shown in Fig. 5-35.

Fig. 5-35. Completed receiver—control end.

Receiver Testing

The receiver can be best checked out by using a standard audio test oscillator or signal generator. Couple the signal directly across the ten-turn link. Adjust the frequency control until a beat note is heard; then peak for maximum signal with the tuning control. The rf tuning is quite broad and can be left alone once it is peaked for the particular transmitter. Transmitter frequency drift is then corrected with the tuning control. Fig. 5-36 shows a husband-wife marine biologist team preparing to use the receiver for tracking of hammerhead and blue shark in the waters off Catalina Island.

Incidentally, the receiver can be coupled to an antenna instead of to the transducer and used for reception of vlf (very low frequency) signals, such as the 60-kHz WWV transmissions and the Navy vlf signals.

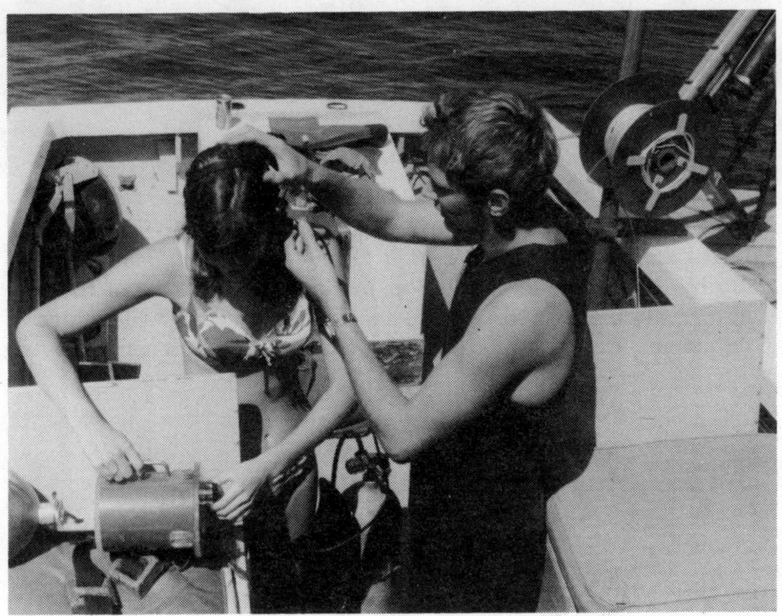

Fig. 5-36. Testing receiver system prior to underwater tracking of sharks.

Transmitter

Next, the transmitter is built. A basic transmitter circuit is shown in Fig. 5-37, while a more powerful push-pull version is shown in Fig. 5-38. These circuits are representative of the type classed as *squegging oscillators.* A squegging oscillator is a blocking oscillator that starts spontaneously, builds up in amplitude, and then stops abruptly. Use of this type of oscillator results in high peak power and long battery life. Layout details of transmitter construction are given in Fig. 5-39.

In both circuits, the pulsing rate is controlled by varying the resistor, RS. Frequency is determined primarily by the inductance of the coil, and secondarily by the reflected impedance of

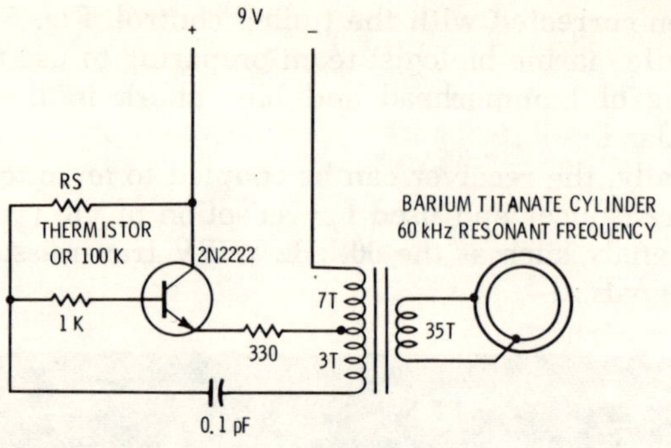

Fig. 5-37. Aquatic telemetry transmitter schematic diagram.

the transducer. The transducer used has a resonant frequency of 60 kHz and exhibits a very high Q, even when loaded mechanically by the castor oil in the housing and the pressure of the seawater. Although it will operate successfully over a wide range of frequencies, the power output drops off rapidly when the driving

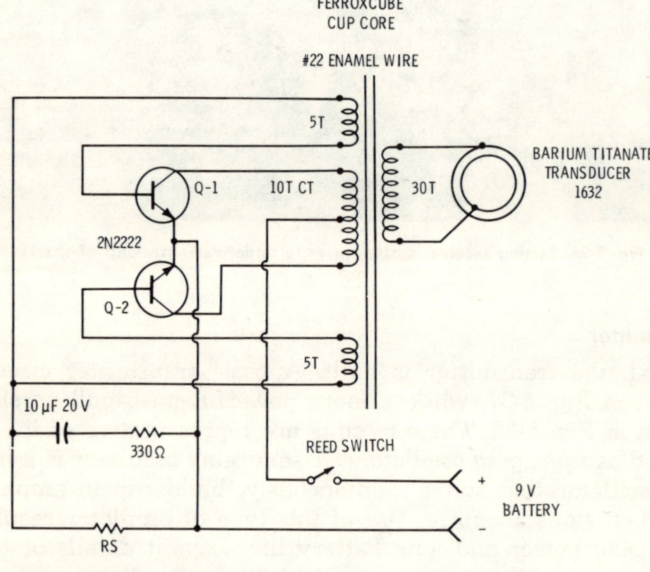

Courtesy E. A. Standora

Fig. 5-38. Push-pull transmitter schematic diagram.

signal frequency is off resonance. Therefore, the coils should be trimmed to match the resonance of the transducer.

Transmitter Testing

After completing the transmitter, apply power and shunt a resistor (a 100K variable resistor is suitable) across the RS terminals. An oscillator can be hooked across the transducer terminals to observe the waveform and to determine if the unit is oscillating properly. The transducer can be placed near the receiver transducer, and the receiver tuned until the clicks are heard in the headphones.

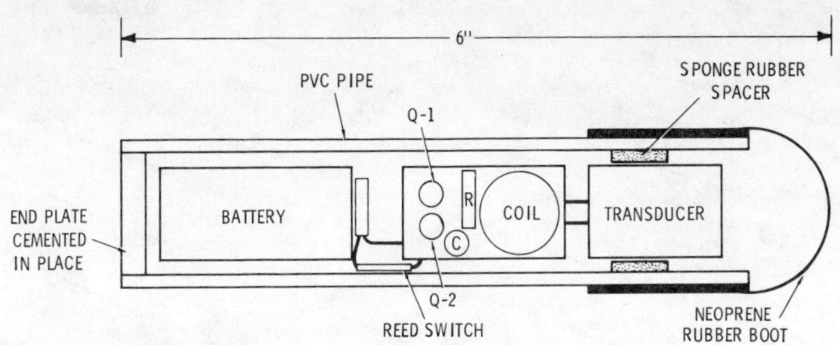

Fig. 5-39. Push-pull transmitter parts layout.

After the unit has been checked out, it can be packaged as shown in Fig. 5-39 for attachment to the marine animal or for experimentation. The range underwater should be in the order of ¼ to ½ mile. Using a thermistor for the sensor resistor, RS, will permit monitoring of water temperature, an important biological parameter. If this is done, the transmitter package should be calibrated over the expected temperature range by using a standard temperature-bath technique.

The real skill in biotelemetry becomes apparent when you mount the package to the animal! For aquatic animals this is usually accomplished by attracting the animal with bait and affixing the package with a small dart. The packages are slightly positive buoyant and can be recovered if a disintegrating magnesium dart is used. These can be made to dissolve in predetermined time periods with reasonably good timing accuracy. The photograph of Fig. 5-40 was taken in the Bahamas, at the Lerner Marine Laboratory.

A good reference book for attaching techniques and special sensor design is Technical Report No. 5, *Development of a Multichannel Ultrasonic Telemetry System for the Study of Shark Behavior at Sea,* by Edward A. Standora et. al, available from the Department of Biology, California State University at Long Beach, Long Beach, California 90840.

A word of warning, check with your local fish and game regulatory agencies before you attach packages to marine animals! Laws may prohibit attachment of packages to certain species.

Courtesy E. A. Standora

Fig. 5-40. A transmitter package attached to a Lemon shark.

PLANKTON SAMPLING NET

The Greek word *plankton* literally means "drifters." Plankton are floating or weakly swimming organisms. Although the term is generally associated with microscopic organisms, it is also applied to any drifting life form. Plankton are divided into two groups: *phytoplankton,* or plants, and *zooplankton,* or animals.

The phytoplankton are the ultimate source of life in the sea. These microscopic plants convert energy from the sun into living matter by photosynthesis. Therefore, phytoplankton production must take place in water of sufficient clarity for light to penetrate.

Zooplankton are the primary animals in the food chain. They feed on phytoplankton, bacteria, and other fine particles of nutrient.

Our knowledge of plankton goes back to 1828 when a British army surgeon, J. Vaughan Thompson, attached a jar to the end of a gauze net in an attempt to collect the early stages of crab life.

Thompson's tow brought in a concentration of microscopic marine life that had never before been seen.

Thompson's net has changed only slightly since its invention. Oceanographers now use nets of varying mesh sizes to filter out the plankton of interest.

The net described here uses a fine mesh cloth, such as sheer nylon curtain material. Fig. 5-41 illustrates the construction details. The net is like a chopped-off ice cream cone, four feet in length.

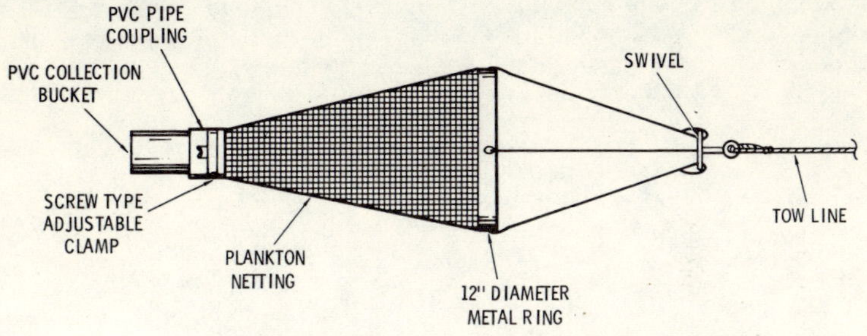

Fig. 5-41. Plankton net construction details.

The net is subjected to considerable strain while being towed. Therefore, it must be carefully reinforced with canvas where it joins the 8-inch ring at the collection bucket end. All seams should be folded double and carefully sewed with polyester thread.

The collection bucket can be made from a plastic jar or, preferably, from a piece of pvc thin-walled pipe which is readily available at most hardware stores. The metal ring for the large opening is made by wrapping steel banding tape around a form, such as a one-gallon ice cream carton, overlapping the ends of the band for several inches, and then wrapping the assembly with waterproof vinyl tape.

The tow line must be securely fastened to the swivel. The swivel assembly is necessary to prevent the line and bridle from twisting during the tow.

In operation, the net is towed slowly over a distance of at least one tenth of a mile. After the net is recovered, the plankton are carefully rinsed into the collection bucket by pouring water through the sides of the net. After the plankton are concentrated in the bucket, the bucket is unscrewed and the plankton poured into a beaker or other container. The collection bucket should be carefully rinsed to make sure that all the plankton have been removed.

The plankton can be observed and classified with a good medium-powered microscope. Place a drop containing plankton on a concave slide, using a dropping pipette or eyedropper. Carefully cover the drop with a cover glass. Do not press down on the cover glass, or the plankton will be damaged. A drop of staining solution may be added to make the plankton more visible.

Bibliography

Bascom, Willard. *Waves and Beaches: The Dynamics of the Ocean Surface.* Garden City: Doubleday & Co., 1964.

Bergaust, Erik, and Foss, William O. *Oceanographers in Action.* New York: G. P. Putnam's Sons, 1968.

Bowditch, Nathaniel. *American Practical Navigator.* H.O. Publication No. 9, Washington, D.C.: U.S. Navy Hydrographic Office, 1962.

Boy Scouts of America. *Oceanography.* Merit Badge Series, New Brunswick: Boy Scouts of America, 1965.

Briggs, Peter. *Men in the Sea.* New York: Simon and Schuster, 1968.

Buehr, Walter. *World Beneath the Waves.* New York: W. W. Norton & Co., 1964.

Bureau of Naval Personnel. *Introduction to Sonar.* Navy Training Course NAVPERS 10130-B, Washington, D.C.: U.S. Government Printing Office, 1968.

Carson, Rachel. *The Edge of the Sea.* New York: Signet Science Library and New American Library of World Literature, Inc., 1955.

———. *The Sea Around Us.* New York: Signet Science Library and New American Library of World Literature, Inc., 1961.

Dean, Anabel. *Exploring and Understanding Oceanography.* Westchester: Benefic Press, 1970.

Dugan, James. *Man Under the Sea.* New York: Macmillan Publishing Co., Inc., 1966.

Fisher, James. *The Wonderful World of the Sea.* Garden City: Doubleday & Co., Inc., 1970.

Gaskell, T. F. *World Beneath the Oceans.* New York: Doubleday & Co., Inc., 1964.

Howe, Mark Wakeman. *Student Syllabus.* Santa Ana: Orange County Department of Education, 1972.

Mackay, R. Stewart. *Bio-Medical Telemetry.* 2nd ed., New York: John Wiley and Sons, Inc., 1970.

Schraff, Robert. *The How and Why Wonder Book of Oceanography.* New

York: Wonder-Treasure Books, Inc., 1964.

Spilhaus, Athelstan F. *Weathercraft*. New York: Viking Press, 1951.

Standora, E. A., et al. *Development of a Multichannel Ultrasonic Telemetry System for the Study of Shark Behavior at Sea*. Technical Report No. 5, Long Beach: California State University, 1972.

Telfer, Dorothy. *Exploring the World of Oceanography*. Chicago: Children's Press, 1968.

Tucker, D. G. and Gazey, D. K. *Applied Underwater Acoustics*. London: Pergamon Press, 1966.

U.S. Coast Guard. *Oceanographic Sensor Study*. Report No. 599009-1A. Washington, D.C.: U.S. Coast Guard, 1970.

U.S. Naval Oceanographic Office. *Questions About the Oceans*. Publication G-13, Washington, D.C.: U.S. Government Printing Office, 1969.

———. *Science and the Sea*. Vol. 2. 3M Document SP-132, Washington, D.C.: U.S. Government Printing Office, 1969.

Urick, R. J. *Principles of Underwater Sound for Engineers*. New York: McGraw Hill, 1967.

Yasso, Warren E. *Oceanography, A Study of Inner Space*. New York: Holt, Rinehart & Winston, 1965.

Index

A

Abyssal deep, 39
Abyssopelagic zone, 39
Acoustic
  energy pulse, 16
  transducer, 75
Acrylic pressure hull, 33
Active sonar, 15
Algae, 74
Anemometer
  calibration of, 71
  construction of, 69-71
ASDIC, 15
Antisubmarine warfare, 10-11, 15
Aquatic
  biotelemetry system, 79-88
  zone, 38, 39
Articulated diving suit, 21

B

Barium titanate, 82
Bathymetry, 37
Bathypelagic zone, 39
Bathythermograph (BT), 18-20
  expendable, 20
  pressure piston, 19-20
Beaufort
  Francis, Sir, 41
  number, 41, 42-43
Behm, Alexander, 16
Benthic zone, 38
Biologists, 7, 9
Biotelemetry system, aquatic, 79-88
Bottle
  Nansen, 11-12
  sampling, 12
Bourden tube, 19
Bowditch, Nathaniel, 10
Buoyancy chamber, variable-, 27

C

Cable controlled underwater recovery vehicle (CURV), 22-25
Calibration
  current-speed meter, 68
  electronic thermometer, 61
  transmitter package, 87
California State University at Long Beach, 88, 92
Cast, BT, 20
Castor oil, 86
Celsius scale, 14
  conversion, 48-49
  developer, 48
Centigrade, 48
Chemists, 7, 9
Color
  comparator kit, 75
  seawater, of, 74-75
Compass, lensatic, 72
Components, electronic, 52, 54
Construction of
  aquatic biotelemetry system, 79-88
  current-speed and direction-sensing system, 61-68
  electronic thermometer, 57-61
  plankton sampling net, 88-90
  Secchi disc, 72-75
  underwater sound system, 75-79
  wind-speed and direction indicators, 68-72
Construction techniques, 51-54

Continental
  shelf, 37-38
  slope, 37-38
Continuous transmission frequency-modulated (ctfm) sonar system, 32
Cook, ship's, 8
Coriolis force, 45
Cousteau, Jacque Yves, 22
Craftsmen and technicians, 7, 9
Cup and vane instruments, 69
Currents
  force
    causing, 45
    coriolis, 45
  major, 45, 46
Current-speed and direction-measuring system, 61-68
  calibration of, 68
  construction of, 61-68
  indicator, 63
  sensors, 62, 63-68

D

Data processing specialist, 7
Deep Quest
  submersible, 31-33
  support ship, 31
Deepview submersible, 33, 34
Depth-sounder, mechanical, 16
Diaphus, PC-14C submersible, 34, 36
Display, sonar, 23
Diving, 20-22
Drifters, 88

E

Echo sounder, 16-18
  definition, 16
  electronic, first, 16
  fishermen, use by, 18
  navigation aid, 17
Electronic
  components, 52
  engineers, 7, 9
  thermometer, 58-61, 68
Engineers
  civil, 9
  electronics
    communications, 9
    data processing, 7, 9
    instrumentation, 9
    space technology, 9
    video, 9
  marine, 9
  mechanical, 9
  photogrammetry, 9
Epipelagic zone, 38
Euphotic zone, 39, 74

F

Facemasks, closed-loop, 31
Farallon Islands, 25
Fetch, 41
Fin, 46
Fish
  -eye lens, 28
  tagging, 80
Fishing floats, 69

G

Gagnan, Emile, 22
Gear, motor vessel, 23, 24, 25
Geologists, 7, 9
Goggles, underwater, 21
Gram, 48
Gravitational forces, 39, 41
Guyot, Arnold, 38
Guyots, 38
Gulf Stream, 10, 45

H

Headset, surplus, 84
Heat sink, 53
High tide line, mean, 37
Homodyne, 81
Hydrophone, 15, 16
  construction, 75-78
  system, testing, 78
Hydrospace, 29, 31
Hydrostatic pressure, 15

I

Iceland, 45
Indicators, wind-speed and direction, 68-72
In situ, 15
Instrumentation, 7

K

Kelvin, Lord, 16
Kelvin scale, 49

L

Laws, transmitter attachment, 88
Le Prier, Yves, Commander, 22
Liquid mercury, 32
Liter, 48
Littoral zone, 39

M

Magnets, 63, 64, 67
Manipulator, hydraulically operated, 23
Manned submersibles, 29, 31-36
Marine
  biotelemetry, 79
  sounds, frequency range, 76
  zones, 38
Mask, face, 21-22
Mathematicians, 7, 9
Maury, Matthew F., Lt., 9-10, 45
Maximum tides, 41, 44

Measurement system
conversions, 48
English, 47, 48
French, 48
United States, 47, 48
Mercury
liquid, 32
thermometer, 57
Mesopelagic zone, 39
Messenger, 12
Meteorologists, 7, 9
Meter, 48
Metric system
unit of:
length, 47-48
mass, 47-48
volume, 47-48
Metron, 48
Mile, nautical, 48
Mineral deposits, deep-water, 28
Motorboating, 78

N

Nansen
bottle, 11-12
Fridtjof, 12
Nautical Exploration Device and Recoverer (NEDAR), 28
Naval Electronics Laboratory, 16
Naval Experimental Manned Observatory (NEMO), 32, 33
Neritic zone, 39
Net, plankton, 88-89
Nine-volt dc power supply, 54-55

O

Ocean currents, 45-46
Ocean outfalls, 34
Oceanic zones, 37-39
Oceanographers, physical, 7, 9
Oceanographic
equipment, construction of, 51
expedition, first major, 10
information, 10
Oceanography
concepts, basic, 37-46
definition, 7
disciplines, 7
Oscillator
audio, 84
blocking, 85
squegging, 85
test, 87
twin-T, 82-83

P

Parts identification, 52-53, 54
Passive sonar, 15-16
"Passport to inner space," 22
PC-14C submersible, 34, 36
Pelagic zone, 38
Period, wave, 39-40
Photic zone, 39
Photosynthesis, 74, 88
Physicists, 7, 9
Phytoplankton, 74, 88
Piston corer, 33
Plankton sampling net, 88-90
Pollution monitoring, 34
Polyvinyl chloride (pvc), 59
Power supply, constructing a, 54-55
Pressure hull
acrylic, 33
steel, 34
Project, simple, 52
Pulse code modulation, 80

Q

Quadrature, 41, 44

R

Radioactive waste, 22, 23
RCV-125, 28-29, 30
Research vehicles, golden age of, 29
Research vessel Gyre, 35, 36
Receiver display, 15, 16-17
Reconnaissance tasks, 34
Recorder, cassette, 81
Reversing thermometer, 12-15, 57
"River in the ocean," 45

S

Sampling
bottle, 11-12
net, plankton, 88-90
Sardinia, sea of, 9
Sargasso sea, 74
Savonius rotor, 64, 68
Scientists, 7, 9
Sea Cliff, research submersible, 36
Search sonar, 28
Seawater
salts, 10
temperature, 57
transparency, 72
Secchi disc, transparency and the, 72-75
Sediment samples, 23, 34
Seismic disturbances, 39
SCUBA, 22
Sewer outfalls, inspection of, 31
Sharks, 80, 85, 87, 92
Shrimp, snapping, 16
Snoopy, 25-27
Soldering, 52-54
Sonar
active, 15
beam scanning, 15
display, 23
passive, 15-16
searchlight, 15
Sonar display, 23
Soundings, 9, 16

Sound
speed of, 18
system, underwater, 75-79
Squegging oscillators, 85
Spilhaus, Athelstean, 18, 92
Standora, Edward A., 87, 89
Stereoscope viewing, 28
Submarine canyons, 37-38
Submarines, work, commercial, 34
Submersible
cable actuated teleoperator (SCAT), 25-28
deep-diving 36
Deepview, 33, 34
manned, 29, 31-36
small, 34
sophisticated, 31-36
tethered, 28
untethered, 33
Subsea pipelines, 34
Supralittoral zone, 39

T

Team, oceanographic, 7, 9
Telemetry, 79
modulation, pulse, code, 80
system
earphone, 84
receiver, 81-82, 84-85
transducer, 82-83
transmitter, 85-87
Telephone-type toroidal coil, 81
Television
cameras
closed-circuit, 23
low-light, 28
Temperature
measurements, 57
scale, 48-49
Terminology, wave, 39-40
Tethered vehicles, remote-controlled, 28
Texas A&M University, 34, 36
Thermistor
current vs temperature, 58
probe, making, 59
response curve, 61
thermometer, use in, 57
Thermocline, locating, 18
Thermometer
auxiliary, 14
bucket, 13
calibration, 14, 60-61
constructing, 58
electronic, 58-61
mercury, 57
pressure gauge, 15
reversing, 13, 14
protected, 14
unprotected, 14
Thompson, J. Vaughn, 88-89
Tidal wave, 39, 40
Tide(s)
celestial bodies and, 41, 44
gauges, automatic, 45
neap, 41
spring, 41
variations, 41, 44
Tool list, 51
Topography, 37-39
Torpedoes, recovery of, 28
Toy motor, 69
Transducer
constructing, 75-78
element, 82
mechanical, 15
sonar, 15-16
Transparency and the Secchi disc, 72-75
TransQuest, support ship, 31
Trochoid, 40
Tsunami, 39
Turtle, research submersible, 35, 36
Twin-T oscillator, 82-83

U

Umbilical cable, 28
Underwater sound system, 75-79
breathing apparatus, 22
sound system, 75-79
television, 26, 28
Unmanned remote-controlled underwater vehicles, 22-29
Untethered submersible, 33

V

Valve, fully automatic, 22
Vane shear device, 33
Variable-reluctance transducer, 75-78
Vlf (very low frequency), 85

W

Water, sound speed, 17, 18
Waterproofing connections, 60
Wave(s)
breaking, point, 41
simplified, 40
speed, 40
tides, and currents, 39-45
Wheatstone
bridge, 58-59
Charles, Sir, 58-59
Wind
speed, 40-43
-speed and direction indicators, 68-72
vane, 71-72

X

Xylene, 19

Z

Zones, marine, 37-39
Zooplankton, 88